✔ KU-708-291

WITHDRAWN

50

Construction · POLYTECHNIC

General editors: Harold Lambeth, FIOB, FCIOB
and W. P. Lansdell, BA.

Timber frame housing — Tub Fletcher
Security on site — Tom Edmond
Glazing — Stanley Thompson
Steel reinforcement — Frank Jackson
Site surveys — Jim Lacey
Site engineering — Roy Murphy

Titles to be published in the Series:
Mixing and placing concretes — Peter Shane and John Baker
Site carpentry and joinery — Keith Lunn
Site planning — Tom Gallagher
Careers in the building industry — Laurie and Paul Allen
Brick, tiles etc. and adhesives — Fred Mann
Careers in brickwork and blockwork — John Broome

WP 04006984

SITE PRACTICE SERIES

General editors: Harold Lansdell, FCIOB, FCIArb, and Win Lansdell, BA

Timber-frame housing – *Jim Burchell*
Security on site – *Len Earnshaw*
Glazing – *Stanley Thompson*
Steel reinforcement – *Tony Trevorrow*
Site safety – *Jim Laney*
Site engineering – *Roy Murphy*

Books to be published in the Series
Making and placing concrete – *Edwin Martin Baker*
Site carpentry and joinery – *Keith Farmer*
Site relations – *Tom Gallagher*
Careers in the building industry – *Chris and Lynne March*
Fixing, fasteners and adhesives – *Paul Marsh*
Exercises in brickwork and blockwork – *Arthur Webster*

Construction site security

LEN EARNSHAW, FIISec

Construction Press

LONDON AND NEW YORK

POLYTECHNIC LIBRARY
WOLVERHAMPTON

ACC. No. 400698

CLASS L

CONTROL

DATE JUN 1987   SITE RS

624
·0289
EAR

Construction Press
an imprint of:
London Group Limited
Longman House, Burnt Mill, Harlow
Essex CM20 2JE, England
Associated companies throughout the world

*Published in the United States of America
by Longman Inc., New York*

© Construction Press 1984

All rights reserved; no part of this publication may be
reproduced, stored in a retrieval system, or transmitted
in any form or by any means, electronic, mechanical,
photocopying, recording, or otherwise, without the
prior written permission of the Publishers.

First published 1984

British Library Cataloguing in Publication Data
Earnshaw, Leonard
    Construction site security. – (Site practice series)
    1. Building sites – Security measures
    I. Title   II. Series
    624   TH375

    ISBN 0-86095-712-8

Set in 10/12 Linotron 202 Bembo
Printed in Hong Kong by ·
Astros Printing Ltd.

Contents

Foreword   vii

Preface   ix

1   The problem   1
     History – Main causes – Conclusion

2   Security considerations and company procedures   6
     Company procedures – Tender stage – Pre-contract stage –
     Security routines – Security incidents

3 ✓ Materials stock controls   10
     Sub-contractors' materials

4   Security at pre-contract planning stage   13

5   Deliveries, acceptance, checkers and checking   16
     Liquid petroleum gas bottles – Discrepancies – Delivery notes –
     Deliveries out of hours – Transfer notes – Sub-contractors'
     materials

6   Physical security on site   23
     Perimeter fences and compounds – Site offices – Locks and
     shutters – Store sheds – Keys – Specialist storage –
     Sub-contractors' materials

7   Out of hours security   34
     Watchman – Police – Alarms – Floodlighting – Guard dogs

8 ✓ Safes, cash, wages, etc.   42
     Safes – Cash – Subbing – Wages

9   Miscellaneous matters and measures   47
     Staff purchases – Off-cuts – Found property – Car parks – Fire
     precautions – Clocking-in offences – Overtime – Trades' rubbish
     – Transfer notes – Telephones – Foreman's diary – Electric and
     gas charges – Powers of search and arrest – Temporary order
     books

10   Damage   55

✓ 11   Security of plant and tools   59
          Plant – Tools

   12   Bonus targets   70
          Payments – Dayworks

   13   Labour relations   72
          Recruitment – Discipline – Dismissals – Redundancy – Employee
          leaving – Absence from work

   14   Sub-contractors   77
          Materials – Plant – Supervision – Payments – Labour only

   15   Liaison with police, fire service, public, schools
          and clubs, and prosecution   81
          Police – Fire service – Schools – Youth clubs – General public –
          New tenants – Empty premises – Thefts or damage – Restitution
          and compensation – Court attendance – Sample letters

   16   Conclusion   93

          Appendices: 96
          1. Anti-vandalism practices in the building industry –
          Preventive measures
          2. Regional Health Authority Notice
          3. Security Code of Practice for Site Managers – Introduction –
          The Cost – Depot – Site security

          Index   105

vi

Foreword

There has long been a need for a comprehensive guide as to the loss prevention consideration applicable to the construction industry.

The author is well qualified to produce this book, which sets out in practical terms the basic areas of loss and the simple counter-measures and procedures open to site management.

He highlights the times that losses are most likely to occur, those materials which are particularly vulnerable to theft/fraud and his experiences in combating such abuses.

It is refreshing to note his acceptance that no matter how thorough site security may be, losses can still occur through either internal or external influences.

It is extremely helpful and appropriate that guidelines concerning authorising prosecution against employees have been simply explained. This is of particular benefit having regard to the complexity of current industrial relations legislation.

He provides the answers in some detail to questions often posed when a thief/vandal is prosecuted to conviction. Restitution of recovered property, reimbursement or compensation to the loser is of particular significance. Other simple Court procedures are explained for the benefit of any potential witness.

The book is written with an earthy style typical of the North Country author.

Although it contains a wealth of information primarily concerned with building sites, it would be invaluable to any person involved in loss prevention.

J. JARVIS (MIISec)
Security Consultant, Construction Security Advisory Service, National Federation of Building Trades Employers

Preface

Although many of the larger building companies now employ experienced security officers to advise site staff on precautions, many firms make little or no provision for prudent foresight in these matters and consequently are very often vulnerable to losses.

It has long been felt, particularly by security advisers, that there was a need for a book to deal with practical, tried and tested, common-sense and economical measures to prevent site losses and to include correct procedures to cover the ordering of materials, acceptance of deliveries, storage, etc. with particular reference to the consequential costs if, due to theft or damage, progress is delayed.

Such 'losses' invariably occur when sites are 'open', and therefore the main need is to direct any such advice to the site and supervisory personnel and anybody who has a responsibility for ensuring profitable results of any building project.

Because of the technical knowledge required, the majority of site managers, agents, general and trades foremen have progressed to that rank by promotion from the natural recruiting area of men 'working on the tools', e.g. bricklayers, joiners, etc. Whilst these people accept that a lack of security can attract problems, in many instances they see their prime concern as supervising the erection of the building to the satisfaction of the client or his architect who has formulated his plans for the building into detailed scale drawings.

Currently, there is adequate provision for the furthering of knowledge of such tradesmen on the 'reading' of drawings, setting out procedures, safety matters, and the interpretation of the Working Rules and the like by attending courses of specific instruction of varying lengths. However, although security courses and seminars are organised in the construction industry, usually of one day's duration, it is regretted that, in the main, they are attended only by security staff and supervisory ranks. Possibly this aspect leads some site staff to the erroneous view that site security is not their responsibility. However, there can be no doubt even though security advisers are employed that the security of a site is the clear responsibility of the site staff.

Security officers/advisers are usually engaged firstly to advise site staff on sensible precautions to take to avoid thefts and damage, and secondly, if any site is violated, to follow up the actions of the police and courts who may have become involved, on questions of restitution and compensation, with the good interests of the employer (company) always being of paramount importance.

Suggestions for anti-theft and anti-vandalism practices, the necessity for security awareness and good housekeeping generally can never be fully appreciated by attending a one-day seminar, no matter how good, and many aspects of security can only be appreciated by the hard and often bitter experiences of years of practice.

Thus, when the publishers presented me with the opportunity to write such a book, it was with some trepidation that I decided to accept the challenge to attempt to fill this need.

The book addresses itself primarily to what site staff can do to ensure the complete delivery and the safe keeping of materials and plant, avoidance of waste and losses due to theft, fire, vandalism and careless damage.

Although it is directed at site managers, supervisors and foremen, it is hoped that they and others responsible for profitable results can read it at their leisure and co-relate some of the experiences of the writer over a period of 15 years, to their own sites and thus, hopefully, restrict thefts and vandalism, etc. to a minimum.

Len Earnshaw

1

The problem

History

The criminal fraternity generally has always realized that building sites were 'fair game' for their operations, mainly because many valuable materials appear to be left lying unguarded about the site, and particularly as up to fairly recent times common-sense security precautions and systems on sites were either lax or non-existent; hence thefts occurred.

Building companies have always accepted that a certain amount of wastage must necessarily result from constructing anything from basic materials, e.g. chipped and broken bricks, off-cuts from timber, electric cable, tiles, paving slabs, etc. and attempts were made to 'cover' this wastage by including an estimated allowance for this anticipated financial loss in the tender price. This amount was generally guesswork, for much depended on the type of site, the actual construction, the materials used, some of which tended to be liable to damage, the type of workmen and the supervision.

In the late 1960s, builders and contractors realized that in addition to the losses by wastage previously mentioned, the estimated losses possibly due to theft and vandalism at that time to the industry throughout the UK were approaching £100 m. per year. This figure was given by the Home Secretary in May 1968, but other authorities even suggested it could be £150 m. The means of calculation were never known, but it was accepted that there was much more thieving, as was common to all walks of life at that time, because of the 'boom' in crime.

So great was the feeling of builders about the increased losses from thefts and vandalism on sites that it was even mooted that an additional small percentage be added to any contract figure to cover these increased losses. This suggestion never got off the ground, for in a 'lowest tender' system of allocating work, everyone was cutting down to a minimum, with any such additions likely to lead to the loss of the contract.

In 1973, following extensive research, the National Federation of Building Trades Employers, as part of their Advisory Service to members, launched CONSEC, which was the code name given to their Construction Security Advisory Service. It was thought that if members were properly advised on precautions many of the losses could be reduced or avoided by good husbandry and house-keeping on sites by making it more difficult for potential thieves to operate.

Incidentally, if one accepts the £100 m. loss figure from the late 1960s, then by allowing for inflation alone, the annual figure in 1982 must be over £400 m., i.e. over £1 m. per day.

Thus there is a problem besetting the construction industry; everybody concerned must be security conscious in an attempt to reduce these losses, and it must be accepted that although care-lessness by buying, surveying and other administrative personnel may contribute to losses, the main causes are found on sites.

The objects of this book and the suggestions are mainly directed at site security where the losses are more obvious and where the professional reputations of the site staff are formed. In many com-panies an annual award is made to the person judged to be the most security-conscious site manager and this idea is commendable and worthy of consideration.

It must be accepted at the outset and without being defeatist in any way, that one will never stop a determined thief and that any security preventive systems will only delay him. Delay and frus-tration are factors thieves do not relish, hence the general reluctance to attack guarded premises, for they accept that the longer the job takes the greater the danger of being caught in the act.

Casual thieves, like customers in supermarkets, may only be tempted if they find materials lying about, easily accessible and transportable.

Bearing in mind these factors, it is proposed to give details of security measures which in all probability will delay a potential thief and in many cases stop him attempting to steal.

However, it is pertinent to observe that it is generally accepted by all responsible senior security personnel in the construction industry that only 10 per cent of losses due to thieving occur during night periods and obviously this factor must have a vital bearing when crime prevention methods are contemplated.

2

Main causes

An analysis of the main causes of losses on construction sites, not necessarily in this order of priority, can be summarized very generally.

Thefts involving employees

There is a large floating labour force in the construction industry and many employees thus become a type of 'casual', in that they start work on a site, e.g. bricklaying, finish that aspect and then move on to another site, most probably for another firm, to do similar work.

The suggestion that many thefts involve employees is not intended to convey that all employees are involved, but with the advance of DIY operations and spare-time jobs (precluded in many regions by the National Working Rules), many employees are tempted and fall. It is not suggested that any system will stop a joiner from taking a few nails, or a plumber a handful of putty, but a serious aspect of all these thefts by employees is that many feel that they are justifiable 'perks' and do not regard it as theft, which undoubtedly it is. Likewise, clerks, office staff and others who take paper, typewriter ribbons, rubbers, carbon paper, pens and the like hardly ever contemplate that this is also straight-forward theft.

This attitude is prevalent on building sites where a type of 'barter' system very often operates. This is best explained by example. A painter wants a piece of glass to repair his greenhouse and asks the glazier for it. He, thinking he is 'helping a workmate', cuts out the required shape from a large sheet of pre-ordered size. To cover up, he promptly smashes the remainder as it is now useless for the place intended. If a query is raised, and sooner or later it must be, vandalism or a similar excuse is given. The loss does not stop there, for on handing over the glass, the glazier, as a token of thanks, asks for a 'drop' of paint to do a little domestic job. Suitable jars, etc. are rarely available on sites, and regrettably the 'drop' becomes a 5-litre can. Such thieving is most difficult to counter and can be very costly to employers.

If any employee is caught stealing, the company procedure must be followed, but it must be stressed that it is very important to have a standard procedure for it would be wrong to prosecute one offender and not another.

3

There are many ways of dealing with such cases, but having regard to the complexities of modern industrial relations law, it is far safer to report the facts to the police with a view to prosecution. Alternatives are immediate dismissal for gross misconduct, suspension or other disciplinary action such as a caution, but if anything other than a prosecution is launched, it can be strongly argued that in effect a licence has been granted for thieving by everybody.

Other losses, attributable indirectly to employees, are often brought about by careless talk in the local hostelry. Experienced thieves can easily glean from an unsuspecting and innocent employee details of deliveries to a site, which stores are most vulnerable, which stores hold the most valuable property, methods of security, and details of any security patrols on sites, etc.

Short deliveries

It is not suggested that suppliers are responsible for these but it is necessary for *all* materials to or from sites to be checked and counted. It is not unknown for part of a load to disappear *en route* to the customer, e.g. ready-mixed concrete, bricks, bags of cement or plaster, but more information will be given in Chapter 5.

Bad site storage

This aspect will be covered in more detail in Chapter 6, but suffice it to observe at this stage that if property is left lying about the site, a trespasser may think that the property has been abandoned, in which case prosecution for theft may be difficult. An essential ingredient in all cases of theft is a guilty state of mind, i.e. that he knew he was stealing, so that if in defence a thief says that he thought the property was abandoned and he is believed, a prosecution must fail.

Thefts by outside agencies

In general terms, almost all security measures introduced on a site are aimed at this category and full details of suggestions and basic necessities will be covered later. A casual approach by relying on insurance claims for such thefts is bad.

Vandalism and damage

There is no such offence as vandalism; it is plain, straightforward, wilful damage whether committed by employees or by trespassers. Other damage, possibly accidental, can be attributed to poor storage on site because of site inefficiency.

Opinions have been advanced, though unfortunately they cannot be substantiated by facts for lack of records, that this is the worst cause of site losses, particularly when projects are nearing completion when most replacements and repairs are either impracticable or very costly.

In current times, regretfully, vandalism appears to be a way of life to some sections of society, and so much concern has been aroused by the subject that in 1975 the Home Office published the findings of a working party of the Home Office Standing Committee on Crime Prevention which had investigated the causes, etc. of vandalism. A special section of that report gives details of suggested 'anti-vandalism practices' in the building industry. This is reproduced in Appendix 1. It is abundantly clear that there is no ready-made quick answer to prevent or restrict vandalism.

Avoidable wastage

While a certain percentage of wastage is accepted, as previously stated, there is no doubt at all that the phenomenal wastage on the majority of building sites can be drastically reduced by sound practical supervision and good site husbandry.

Conclusion

While it is generally accepted that there is a serious problem of losses in the building industry, there is no doubt that it can be reduced considerably by simple, common-sense economic precautions. With this object in view, many building companies now have specialist security advisers on their staff to advise management on methods of security systems and loss prevention generally.

However, before dealing with security systems in detail, one must analyse the time and occasions when security must be considered and specific action taken.

2

Security considerations and company procedures

Having accepted that there are serious losses in the construction industry from the many causes already mentioned, it is an absolute necessity that some positive action or system be adopted by every builder to reduce these losses to a minimum.

It is a well-known fact that companies who have taken security seriously by introducing procedures and educating their staff have never regretted nor abandoned the policy.

Damage, theft and wastage can be prevented by careful thought and appreciation, taking precautionary action and by physical means, such as watchmen, alarms, etc. However, it must be accepted that on economic grounds alone it is impossible to make all stores into fortresses, for the very temporary nature of those on building sites does not help. Furthermore, while determined thieves are difficult to counteract, lack of sensible and elementary precautions must not make a trespasser into a thief by exposing him to temptation.

Before highlighting detailed security measures which all companies are advised to explore, one must first analyse the various phases in a building project when consideration is given to security aspects.

Company procedures

All companies have recognized general procedures (often reduced into print as Standing Instructions given to all relevant staff) to cover the many aspects of a building programme and, while they are not always considered as such, many are undoubtedly aspects of security.

Examples of these procedures are: the ordering and acceptance of materials; the requisitioning of plant (its delivery and return are two aspects which are dealt with more fully in Chs 5 and 11 respec-

6

tively); the necessity and manner of timekeeping of labour; detailed measuring, etc. of satisfactory work performed on site so that payments made to sub-contractors, etc. are as accurate as possible; the documentary proof required to substantiate claims for expenses incurred by staff and many other aspects of administrative control of outgoings from the company, e.g. the supply of petrol for company cars which, unless otherwise previously agreed and contracted, should be restricted to official mileage. Reference could be made to many more security aspects which are invariably controlled by company procedures, and all contractors must realize that if their organization, administration and security measures are so slack that even an honest man might be tempted to steal then their system is at fault.

Tender stage

Before a tender for a building contract is completed, or before the price for houses, etc. built for sale is fixed, many aspects of probable expense are explored, e.g. the availability and possible costs of importing labour, etc. to the district. Additionally, the anticipated costs of necessary compounds, lighting, watchmen, obviously affected by the location of the project, must also be assessed and must be included in the tender figure.

It is often advanced as an excuse, particularly after a major theft from a site, that a necessary compound or similar security precaution was not established because there was not sufficient money in the contract figure to cover it. While one must accept that sometimes diligent enquiry and research reveal that such security measures could be kept to a minimum, it is more than likely that against all pointers, management has taken a risk in not including a sensible figure for a security contingency in the tender figure.

It is appreciated that if excessive caution is applied by including a high cost for security a tender may be lost, but it is very foolish to ignore completely possible security costs, for the consequences may result in a project having a very low profit or even a loss.

Thus it is imperative that at the tender stage security must be considered and any expected costs included in the tender figure.

In this connection some builders have been known to boast that although many of their sites do not have any security they 'get away with it', and this type of builder often argues that the cost of a system of security could be far greater than any thefts which might take place. This is a fallacy, for one can never ascertain the

7

effects of a security system nor how many thefts have been prevented, nor if it becomes well known among thieves that a sife has no security, the adverse effects that can have.

Up to a few years ago, there was an erroneous attitude in some construction companies that money spent on common-sense precautions was a straightforward loss, without any thought of the loss which may be suffered if precautions have not been taken. Fortunately, this attitude is gradually changing.

Pre-contract stage

Once the job is obtained then there must be detailed planning by everybody concerned on the general security set-up and procedures to be adopted on the site, but this aspect is covered in detail in Chapter 4.

Security routines

It is not unusual during the progress of a building project for the existing security measures to be reassessed in the light of current events. It may well be that due to longer evenings or vice versa or a spate of damage, consideration will have to be given to varying the hours of a part-time static guard or to transfer existing or install new floodlighting on a particularly vulnerable area of the site, e.g. finished blocks, etc. Obviously the site foreman/manager should discuss these possible changes with the relevant personnel, such as the security officer. Any security procedure or system should not be so rigid that it cannot be varied or changed if necessary to counteract current happenings.

Security incidents

If an incident occurs such as a break-in, a theft or extensive damage, the site manager must have clear specific instructions, generally included in the Standing Instructions of all security-conscious companies, on the action he must take, e.g. calling police, notifying supervising staff, submitting insurance claims, etc. Advice on such general procedures is covered more fully in Chapter 15.

There are many more minor incidents which can occur during a building programme which are not always considered under the security heading, but undoubtedly any avoidable losses to the employer whether caused by carelessness, bad supervision and the like are security matters.

Site management must appreciate that even working on a 5 per cent net profit base (not very high) a £100 loss, no matter how caused, equals the net profit on £2000 worth of work. They should also realize the losses to the company if this method of calculation is correlated to the cost of punctures to site transport by nails left lying about the site, or the possible maintenance costs to new wooden floors which have been soddened by plasterers who have been permitted to mix their materials on them, albeit in receptacles, or the maintenance costs to the builder when rectifying dampness caused by careless dropping of mortar into wall cavities.

Very often a simple theft or damage can cause delays of varying lengths, possibly delay of the hand-over date, say for example, the loss of the warden's master key for a sheltered accommodation block, or the damage to specially made door handles for a prestige contract, e.g. a civic centre. Apart from the annoyance caused by such examples, site managers must appreciate that the cost of the project is governed, among other things. by the length of time it will take to complete. Any delay automatically increases the cost of the job by additional overheads, extra insurance for the period, watching costs, expense of retention of office accommodation and storage sheds on site plus, in many cases, compensation payments in accordance with any penalty clauses for running over the contract period, and last, but by no means least, the loss of goodwill of a client by failing to meet the completion date.

These 'tangent' losses are not always fully appreciated by site staff, who must constantly endeavour to complete the work in the time laid down in the 'programme'.

It therefore follows that all site managers must appreciate that adequate site security measures to protect vital materials, etc. are a necessity.

3

Materials stock controls

If materials are on site too early then they become an additional security and storage hazard, yet if they are too late on site, then the project is held up.

All companies assess how long the project will take to complete and the times of various stages of the construction. This is called the 'programme' and every aspect of the job is based on it, i.e. labour required, type and numbers.

Equally, this applies to materials which, in theory, are scheduled to arrive on site just before they are required. With this in mind, whether the materials are delivered direct to a site or to a depot for onward transmission, when orders are placed with suppliers, firm delivery dates are made and the copy orders sent to site contain details of these delivery dates.

For materials such as bricks, aggregates and cement, the need for which can only be ascertained from day to day, the company will usually place a covering order with the suppliers and authorize the site management to 'call forward' or telephone for delivery as required. In such instances the overall requirements for the site must be stated in the bulk order and the manager must keep a running, reducing total on site so that he can readily assess his future requirements against the order.

Without materials a site must run to a halt, therefore a site manager must be constantly assessing his supplies. On the other hand, if a site is behind programme, and materials are scheduled for delivery well before the time when they can be used, it is the duty of site management to be aware of this fact, and, if possible, to postpone delivery to a more convenient date. It is appreciated that this is not always as easy as it seems, for costs etc. agreed by the buyers may be affected and extra expense incurred. Site management must then decide whether to accept delivery with the probability of additional storage on site, if space permits, or whether to accept delivery on the scheduled date at the depot with the consequent additional handling costs for site delivery by the company's own transport. A practical point for consideration in this latter case is whether the company's transport is suitable and adequate for such a delivery, e.g. large pre-double-glazed windows, normally delivered by suppliers in specially adapted vehicles, could not, with safety, be transported on flat wagons.

If a site is behind programme, site management must be on top of the situation so far as acceptance of material deliveries is concerned. Consider, for example, a house-building project which is running five weeks behind programme, where the schedule is the 'hand-over' of five dwellings a week. Each dwelling has two porcelain toilet basins, so it is easy to calculate that an extra fifty or more would be on site for five weeks, thus causing a security and storage hazard and, just as seriously, the possibility of the material being 'chipped' or damaged. It follows that WCs, baths and kitchen fitments would all add to this problem.

Sub-contractors' materials

The previous comments are applicable to the company's own materials, but very often confusion, frustration and straightforward clashes occur with materials of sub-contractors, and even though the sub-contract documents endeavour to cover all such aspects, a good site agent, in the interests of his employer, will always ascertain at the outset the answers to the following questions:

1. Who provides the store for these materials?
2. Who has keys and who has access to these stores?
3. Who receives them on site? If the sub-contractor is on site, this is resolved, but if not?
4. Who checks the correctness of the delivery, particularly special materials when the builder's men are not always competent to decide?
5. Who pays for the off-loading?
6. Who pays for delivery from store to scene of operations?
7. Who is responsible for the security of these materials?

It is generally accepted that under the standard form of contract in use any losses due to theft or damage are the sub-contractor's responsibility until fixed, then it is the main contractor's liability. There is a grey area about what is meant by 'fixed', but in the case of the completion of plumbing and electrical first fix, this is usually accepted as 'fixed' as opposed to the final completion of the installation.

Checkers and checking are covered under Chapter 5.

4

Security at pre-contract planning stage

Security is not just the provision of good locks, bolts, bars and the like, but is more far reaching, for there are many aspects which must be planned before the job starts to eliminate the possibility of avoidable expenses being incurred against the builder.

Before any contract is started, all the personnel affected should always meet to discuss the project in detail and it is at that stage that the security of the site must be considered.

Many points are considered at the pre-tender stage such as availability of labour, tipping facilities, nearness of gas and electric services, but by no means least the reputation of the area for thieving, vandalism and the like must be explored. A word with the local Crime Prevention Officer may elicit knowledge on this score and if the information about the locality is bad then the depth and quality of the security measures to be taken and the estimated cost involved must be allowed for in any tender. All the information that can be obtained about possible site conditions is a vital matter in the field of security.

The type of district, e.g. town centre, demolition site, green field, etc. has a vital bearing on the security measures to be adopted; for example, if in a town centre it may be advisable to seal off the site completely with all offices and stores inside one large compound with restricted access. On the other hand, an open field site would possibly be too large to encircle, but a materials compound could be a necessity. In this case, the location of the compound must be identified and this can be affected by the order of building, for the compound should be easily accessible at all stages of the project. Ideally it is advantageous to have the compound centrally situated on site, but otherwise building should start away from the compound and progress towards it. In this case the site office complex, usually near to the compound, will provide an overall view of the site, which would not be so if the building were started and worked away from the compound.

Thought must be given to the question of the location of, and entrances to, the compounds, for once erected, often at great cost, fences should not be violated by company employees to create an additional, possibly temporary access. Once the perimeter has been interfered with, its value is lowered. *Always remember that a compound is only as strong as its weakest link.* One could give many instances where large amounts of money spent on compounds have been a sheer waste because of subsequent thoughtlessness.

Details of the types of site stores are explored in Chapter 6, but in very general terms timber, wood and corrugated sheets and even metal stores can become suspect after constant usage and therefore at the pre-planning stage, when the order of erection is considered, the question of building garages, bungalows and other similar buildings already in the scheme as a priority for use as stores is advocated. The use of temporary doors for the garages, as opposed to the metal type which can easily be damaged, and temporary doors and boarding up the windows of bungalows is recommended. Experience has shown that such adapted stores are stronger and more satisfactory and drier than temporary wooden store sheds. Obviously, if these garages and bungalows are to be plastered out, this operation is left to a later date. In any case, such adaptations will only be made if the premises are scheduled for handing over at a later stage.

It is at this pre-start date that it is always advisable to get in touch with the local police, the Crime Prevention Officer, who will welcome a discussion particularly about the locality, types of thief, proneness to vandalism and possibly local labour. He can possibly tell of experiences of previous builders who have operated in that area, their problems, possibly solutions and of course the type of security they adopted and whether it was effective. The means of such liaison. initially by letter, is given in Chapter 15.

Obviously, before any project is started, the site must be visited by the site manager, so that he is familiar with the landmarks shown on the drawing(s) of the site layout. He can also see on his visit if the site has been used as a 'short cut', a play area or any similar hazard, which must be stopped before operations commence. It may be necessary to have notices posted around the site, sometimes in foreign languages, to cater for certain sections of the local community and also publicized in the local press. Evidence of such actions can be of vital importance in the event of a trespasser being injured, etc. after the site works have commenced.

It is not unknown for claims to be made against builders from the owners of adjacent property or from the local authority for damage alleged to have been caused by the builder during his operations, e.g. broken concrete pavings or kerb edgings, cracked inspection covers, etc. said to have been caused by the builder's plant or transport running over them. While it is accepted that any such allegations have to be proved, it is very advisable that if any such damage is seen *before the operations start* to make a written record and even in some cases to have it photographed. Incidentally, any close-up of a broken paving slab, etc. must include some visual evidence of the location to be of any value.

In this context it is worthy of mention that if any vehicles of sub-contractors, materials suppliers or even removal vans bringing in new tenants are seen to do any damage including that previously outlined, the driver's attention must be alerted and his details, including his vehicle number, must be recorded for reference.

5

Deliveries – acceptance – checkers and checking

The maxim that must always apply is that *a signature is only given for materials or property of the correct number and quality actually received.* (Some firms have an additional procedure, established when the order is placed, for the delivery docket, even though signed, to be taken into the site office to be rubber stamped.) This is common sense and all should realize that if a supplier has a signature it is virtually impossible to prove that the goods have not been received, or that in some cases they are of inferior quality.

There are certain rules which must be adopted on all sites to ensure satisfactory delivery and acceptance of materials.

Firstly, the checker must have the intelligence and education to ensure that he is capable of satisfactory work. The 'one, two, three, a lot' type must never be used, neither is it a job for a part-time employee, e.g. a student on holiday from college.

He must be a trusted employee, not prone to the acceptance of 'tips', which can have repercussions, and be able to write his signature legibly, for many disputes have never been settled satisfactorily because of the inability to identify and trace the checker.

If a checker is not used, then only named employees of the correct calibre must be so authorized.

Sub-contractors' employees must *never* be allowed to sign for other than their own supplies. To allow otherwise is courting trouble. For example, if a sub-contractor is engaged on road-making, and the materials such as stone and hardcore are covered by a bulk order (he being paid for the quantity placed) it is in his interests to 'accept' more than is actually received, because once the materials have been tipped, rolled or compacted, etc. it is well-nigh impossible to dispute the actual quantities of each delivery.

Bulk orders, with a call forward mandate with the site manager, can be very risky, but this can be countered if *the checker insists on getting a delivery note with each delivery.* This shows, if bought by weight, the tare, total and load weight with the name of the driver and the number of the vehicle. No matter how many deliveries are made, or how many vehicles used, the rule of *a ticket at the time of each delivery* must never be allowed to lapse, for very often, as the lorries are hired, the owner being paid at a rate per load, and the driver a bonus for extra loads, it is in their interests to get as many loads as possible. The practice, which applies far too often, to accept the loads as they are tipped off, and to sign several delivery notes at the end of a day, is wrong and should never be allowed.

By law, the tare weight of a vehicle must be shown on the near-side chassis of the vehicle. Usually it is painted on, but checkers should be suspicious of tare weights which are only chalked on or which are shown on a removable metal plate screwed to the chassis.

There are many fiddles involving 'adjusting' tare weights, such as the driver and/or passenger not sitting in the cab when being weighed empty, but sitting in the cab when the full load is weighed thus getting an inflated load weight. Similar fiddles are worked with spare wheels, tarpaulin sheets and other equipment which can easily be moved. Each individual item may not appear to have a significant effect, but collectively they can have a devastating effect on the requests for payments made on a builder, particularly if large quantities or high-value materials are being received.

There is a legal provision that if a checker suspects any weight fiddle or discrepancy, he can demand that the driver should take his vehicle, loaded or empty, to the nearest weighbridge for check-ing, and failure to do so is a criminal offence. Site managers should regularly take advantage of this provision if only for the psycho-logical effect it can have on any potential 'fiddler'. If a reputation for keenness by management is created it has the desired preventive repercussion.

If the deliveries are based on a cubic measurement, it is as well to know that all wagons have a designated cubic capacity shown on the nearside of the body and an experienced checker can estimate whether heaped-up loads, if levelled, would fill the wagon and therefore the cube. When several wagons are used, the checker, if efficient, soon identifies the capacity of a particular wagon.

If there is any doubt it is easy to measure the length, breadth and height of the wagon body to get the cubage.

It is relevant to mention that, particularly with bulk orders for building sand, grit sand, clean hardcore, topsoil, etc. if the checker is not qualified or mandated to judge the quality of such loads, he must not allow it to be tipped until approved by site management, for once tipped an acceptance is implied.

It is the duty of the checker when a bulk load of any type of sand, etc. is delivered to indicate the point of tipping on the site and to notify a fellow workmate, generally a foreman, so that the tipping can be checked. To attempt to avoid wastage and more particularly pollution with soil, rubbish, etc. site management should have previously arranged for a concrete, timber or polythene base. The same applies to bulk deliveries of ready-mixed screeding materials.

Delivery of ready-mixed concrete often causes some concern, particularly if it is to be used for over site or sewer covering, etc. In the former case the site management can ensure that the area to be filled is measured up before delivery and if the circumstances are suspicious, the services of a quantity surveyor may be used. However, it is not so easy to check amounts used for fence posts, drain covering, manhole settings, etc.

It is not suggested that it occurs regularly but it is not unknown for the odd cube metre to be 'dropped off' at some unauthorized stop before delivery, and the golden guideline is that if site management makes an obvious determined effort to check a delivery, it may forestall any attempt at short measure.

Weighing a ready-mixed concrete vehicle does not help to obtain the cubic content, but if doubts prevail over a period, the assistance of the local inspector from the Offices of Fair Trading may prove fruitful.

Site managers should be aware that many ready-mixed concrete lorries, even though they bear the name of the supplier, are privately owned by the drivers who are paid on a cube metre/mileage basis. It is, therefore, in the interests of the drivers to get as big a cubage as possible, and cases are not unknown where there has been collusion between drivers and batching plant staff to the disadvantage of the purchaser.

These wagons run to a very tight schedule for obvious reasons, and therefore every effort must be made by site management for a quick 'turn-round' by arranging access as near as possible to the tipping point, for the waiting time charges can be very large and are a straightforward irrecoverable loss.

On some large sites fuel oil is delivered in bulk direct from a tanker into the site store tank and site managers must be alert to the possible fiddles which are often operated. The stock in the store tank must always be taken before the acceptance of a delivery, either by the indicator if there is one fitted, or by dipstick. The tanker driver must also dip his tank and show his dipstick level to the receiver. It is then paramount to note that the supply is being drawn from the tank which has been dipped and after the delivery that the end dip is taken in the same tank and that the dipstick is fully lowered, i.e. the handle is flat against the neck of the tank and not resting on the operator's fingers. If the latter trick is operated the dip of a partially full tank will reveal empty.

Before accepting and signing for the delivery the recipient must again check his own store tank for confirmation that the correct amount has been received.

Opportunity is taken to refer to packing cases, pallets, etc. which are brought to site to prevent possible damage to materials. Many suppliers are now getting more cost conscious and charge a deposit on such cases, etc. and because this fact will be shown on the delivery note, the checker must ensure that the case is there and not damaged. Furthermore, it is good practice to emphasize this extra charge when the delivery note is forwarded to the office, and obviously great care must be exercised in protecting the case, etc. for subsequent return to the supplier.

Since this charging system for deposits was introduced, a new type of theft has become very prevalent by unknown drivers who call on sites, alleging an authority to collect packing cases, etc. with a promise of credit of the amount of the deposit. Many site agents have been 'conned' by this story which should never be listened to unless the driver can produce written authority for his collection, and even so he must sign a receipt which contains particulars of his work address, number of his lorry, etc.

Liquid petroleum gas bottles

It is pertinent in the context of returnable pallets, cases, etc. to refer to liquefied petroleum gas bottles such as Calor, Shell, B.O.C., etc., for the charges for the alleged loss or the late return of bottles cause almost every building company worry and irrecoverable costs.

Initially, site managers must realize that in addition to the returnable deposit on each bottle there will undoubtedly be a variable demurrage charge. This is a charge made after a specified time for emptying the bottle, e.g. Calor allow two months, referred to as the free loan period, for the return of the empty bottle. It is pertinent to point out that this demurrage charge is fifteen times greater per month in the winter months when the empties are more urgently required. At the present time the demurrage charge per month, after the two months' free loan, is 10p per cylinder from April to September, but from October to March it is £1.50. The cylinders always remain the property of the gas company, and if any are lost, the current charge is £16 per cylinder irrespective of size.

Liquid gases are absolutely essential on many sites for heating and lighting, but if only small quantities are used supplies are usually obtained through a local agent, when initially a deposit on the first bottles supplied is invoiced and the hire charge is raised. Further supplies are often restricted to the quantity covered by the empty bottles returned. This method is an expensive exercise.

However, builders who use large quantities generally deal direct with the gas company, who deliver new stock in bulk to the builders' depot, any empties being returned and credited at that time.

Generally, the distribution and collections from sites is the responsibility of the builder, and it is because of this acceptance of detailed delivery that the builder is usually able to negotiate advantageous terms with the gas company. These may include a reduced price for the gas because of the quantity bought, possibly a reduction in the deposit charges, but there is always a clause in the contract allowing, say, two months' free use, after which time an automatic demurrage charge for non-returns on a time/charge basis. There is always provision for substantial additional rebates for a quick return of bottles.

Basically, and in theory, it is a very straightforward system, but in practice all sorts of pitfalls exist as many builders know to their cost, and these are often caused by slackness on sites, possibly because site managers do not fully appreciate the details of the free loan period and demurrage charges, or by faulty office administration. It is a fact that at some time most building companies have experienced difficulty in reconciling the gas company's record with the recorded stock they hold.

It could well be that because they are extensively used in caravans, etc. that many gas bottles are stolen from sites or are 'lost' because site managers wrongly allow the use of their gas by subcontractors on site.

Therefore, because of the possible risk of large losses, all efficient building companies should have a system of accurate weekly returns from sites giving stock at the beginning of the week, supplies received, bottles returned and final balance. This will enable top management to be kept aware of the position at any given time, but it is essential that the correlation of head office records must be given to a responsible person, for the possible financial losses can be very large.

The message should be clear to all site managers that empty gas bottles must be returned as soon as possible. For storage of gas bottles on site, see Chapter 9 under Fire precautions.

On sites where large quantities of gas are used it is not unusual for the gas company to supply and fix pressurized tank on site and supplies are delivered by tanker. To check that the invoiced amount is being received it is essential to check the meter readings on the tank and on the delivery tanker, before and after the delivery.

Discrepancies

The only way to check a delivery, if it is by weight, is by the weigh note which must be carried and handed over, but if it is by number, then by counting. Packages of bricks vary in size and a close scrutiny is necessary as with loose bricks, but all other items must be thoroughly checked against the delivery note and any shortages written legibly on the office and the deliverer's copies. It is useless doing the former and not the latter.

Sometimes, in the case of small items, such as door and window furniture, screws, switches, plugs, etc. deliveries are made in boxes such as tea chests and invariably the driver is in a hurry and requests a signature as soon as possible. Irksome it may seem to him, but the checker should 'check' as is his duty. Giving a signature and an endorsement 'unchecked' is most unsatisfactory. If, however, such an incident occurs, then a physical check at the first opportunity should be made and any shortage reported by telephone and confirmed by letter.

Delivery notes

Each company has its own system, but common to all is the fact that all delivery notes must be forwarded to the local office as soon as practicable, more particularly if a discrepancy is revealed. These notes are vital documents for checking against invoices received from the suppliers and are an important link in any system. Arguments often arise because of the absence of a delivery note, and much embarrassment results when the supplier can produce a signed copy delivery note to substantiate his invoice which is being queried.

Deliveries out of hours

Out of hours deliveries to sites should never be allowed and this point must be made abundantly clear in writing to suppliers at the time the order is placed. However, as lorry breakdowns cannot always be foreseen, with the consequent last minute arrival of materials on site, extra charges in the way of overtime for off-loading can justifiably be claimed from the supplier if he has been so notified.

If a delivery lorry arrives at the site after closure and deposits the goods, usually a small item, near the checker's office a *fait accompli* situation arises.

In such cases the goods must be very closely examined before signing a delivery note, which when returned must have an accompanying letter stressing that this practice will not be tolerated and if any defects are subsequently found, the supplier will be liable for any costs, etc.

Transfer notes

It is equally important when materials are removed from a site, e.g. by returning them to the depot, transferring them to another site or returning them to a supplier, etc. that the transaction is recorded on a transfer note so that the appropriate cost credits and debits can be recorded.

Some companies frown on this transfer note procedure in the belief that the additional administration costs more than the materials concerned, but in the context of tight security systems, transfer notes are a necessity.

Sub-contractors' materials

On large sites there is inevitably a sub-contractors' representative, in which case he is responsible for accepting his own materials, but complications and confusion can arise if the checker on site accepts sub-contractors' materials. He can certainly check numerically, but for quality and other correctness he may not be qualified. This situation should never arise, for prior to the job starting all such aspects, together with the question of who pays for off-loading, storage, etc. should already have been explored and covered.

22

6

Physical security on site

It would be ideal if all building sites were situated on level ground, but such is not the case and, so far as compounds, location of office and stores huts are concerned, the most level piece of ground should be found. A word of caution here, for it is not unknown for huts, etc. to be erected and used for some time, only for it to be realized later that they are on the direct run for a surface drain, electric, gas or British Telecom line. Site managers must bear all these factors in mind when fixing the position and erecting the sheds, etc.

If the huts are not on level ground but are levelled up with wooden sleepers or stacks of bricks, then the 'open' underside of the hut must be properly and effectively closed, for it is on record that thieves have entered wooden huts via the floor. This is far more important with the corrugated metal type of store with no floor provided. In such cases, the sides must be well and truly fixed to the ground, possibly by a road-pin type of fastening, but much more important, access by digging underneath the walls must be made virtually impossible.

Following these general comments, one must detail the security measures which must be observed in specific instances.

Perimeter fences and compounds

There are many types of fences for compounds, namely close boarded, paling, cranked (faced outwards) concrete posts and wire, scaffold pole and wire, even sheet metal, but clearly the type to be used is often governed by local conditions, and on information gleaned from the local police, who have knowledge of previous experiences by builders in their area.

If a close-boarded type is used it is recommended that 'peep-holes' be provided, in theory for sightseers to watch the work in progress, but in practice so that thieves can be seen during non-working hours by patrolling police and the public. If such peep-holes are not provided, one is merely providing an adequate screen for thieves and vandals to operate. It is obvious that these peep-holes must be either small enough to prevent climbing through, or if not so, covered with a bolted wire grill, nuts on the inside.

Close-boarded fences are very popular, for if care is used when dismantling they can easily be transferred to another site. They are ideal for segregating the work from the public when part of a pub-lic pavement is taken over temporarily. It is customary for protec-tion and publicity purposes to paint these fences in company colours and superimpose the name.

All perimeter and compound fences should be at least 2.5 metres (8 ft) high and topped with barbed wire, a further deterrent to a climber.

It is not advisable to utilize existing buildings as part of a com-pound fence for they may afford easy access and any windows or doors in the building are vulnerable. Consequently the building itself must be made secure, otherwise there could be a weakness in the compound.

Entrances to the site or compound must be as few as possible, their location having been decided at the pre-planning meeting. One entrance must be adjacent to the checker's office, which must be suitably identified by a notice.

There must also be a notice prominently displayed on the per-imeter fence or compound that all visitors must report to the site office, the location of which must be indicated.

Only in the direst emergency must temporary entrances be made, and then properly and securely sealed afterwards as near as possible to the original state.

Gates to fences and compounds must be substantial and incapable of being lifted off. If crook and band hinges are used, then the top crook must face downwards.

All gates and compound perimeter fences must be close enough to the ground to prevent crawlers, particularly children, from gain-ing access. This is very important if dogs are used in the com-pound. The use of dogs is covered later in Chapter 7.

To prevent easy cutting or forcing the lock by inserting a steel rod, etc. good-quality padlocks of the hardened-steel close-shackel type and good-quality chains to secure gates, etc. must be used. It is a waste of money to use an expensive lock if the chain can easily be cut, and the opposite equally applies.

One golden rule which must never be broken is that *in no circumstances must materials be stacked against the inside or outside of a perimeter or compound fence.* The reason is obviously that such action would enable thieves to climb into or out of the compound with ease and thus render the compound useless as a security measure. Equally, if scaffolding or like material is used for timber storage shelving, either inside or outside, it must be far enough away from the fence to discourage a determined thief from using it as a jumping-off platform into the compound.

It is worth emphasizing that a security fence is only as strong as its weakest link, and if breakages to the fence are made either accidentally or otherwise, unless that break is securely repaired the whole initial expense of erecting the compound has been wasted.

Site offices

These must be clearly marked so that all visitors know where to report. There is always the possibility that persons allegedly seeking employment are really 'casing' the site by trespassing around and when challenged give the excuse of searching for the site offices.

Sectional wooden site offices are the most common but permanent huts of the 'Portakabin' type are available either for purchase or hire, but whatever type is used they must *not* form part of the perimeter or compound fence for if an entry is effected into the office it is then very easy to continue into the compound. If several cabins are used either for offices or stores, they should be sited at least 1–1.5 metres (3–5 ft) apart because of the fire hazard.

The checker's office, suitably marked, must be near the site entrance and preferably elevated so that checks can more readily be made. Furthermore, this height can have a psychological advantage against drivers of vehicles with high cabs who may try to intimidate the checker (see Ch. 5 for checking delivery notes, transfer notes, etc.).

Locks and shutters

All site offices must have adequate locking-up facilities with as few named key-holders as possible and all windows must be fitted with shutters, which must be closed during non-working hours.

Only the best type of good-quality mortise locks should be used on site offices and the keep can be strengthened by fixing a metal strip on the inside of the door jamb, for it is well known that physical force need not be excessive to split the wooden keep or jamb in an average site office.

If rim-type locks are used, then a metal gaiter can be fixed around the metal keep for additional strength.

Sometimes hasp and staple-type fastenings are used on office and store cabins, and it is very important that they are properly fixed, i.e. no screws should be left showing when in the locked position. It is no use paying out for an expensive locking system if this can be bypassed by unscrewing the fixing screws. Close-shackle padlocks, i.e. those where it is impossible to force the lock by inserting a piece of metal under the shackle, should always be used in a 'hasp and staple' situation, and the fixing of a bolt, nut on the inside, through the hasp is an added security measure.

Shutters, preferably wooden, which are fitted must be securely closed and fixed with bolts, nuts on the inside, and all screws and hinges on the shutters must be burred to prevent opening by a screwdriver. This latter precaution also applies to hasp and staple fittings where it is impossible to cover up the screw heads.

Small-mesh permanent shutters are very effective and avoid the nightly ritual of fastening as with wooden shutters, but they must also be secured from the inside.

Store sheds

All the suggestions about level ground, locking facilities and shutters apply equally to store sheds, but additional thought must be given to using the best available store sheds for the most vulnerable or expensive materials.

If sheds are of the sectional type, the sections of which are held together with bolts, then it is common sense that all nuts are on the inside. Great care must also be applied to check that the roof is properly secured for it is not unknown for roofs to be blown off in a gale, and it is a well-known fact that thieves always try the roof first for it is rarely 'alarmed' up if there is such an installation. This is particularly applicable to sectionalized metal-type stores regularly used for plumbing goods where very often roofs are held down by twisted pieces of wire on to the side. Furthermore, it is not unknown for such pieces of wire to be visible from the outside, giving easy access for cutting, etc. If it is considered necessary because very expensive goods are being stored to have bars on the inside of window openings (wrought-iron mesh is often fitted) then all fixings must be on the inside of the store.

All site management must ensure by constant checking that even during working hours, all site offices and stores are always secured when unattended. This not only deters site 'visitors' but all site employees from being tempted.

In this context it is worthy of mention that workmen must never be allowed to collect their materials from a store, but the materials must be issued by a responsible storekeeper (it is accepted that on a small site this may be the general foreman) and that any such issuing store must have a counter barrier to stop workmen from wandering around the store at their leisure. Only sufficient materials for that day's work should be issued; it is fatal to issue materials well in advance for this tempts workmen to 'take it home for safety' as opposed to leaving it on site where it may be stolen. Store keys must never be handed over to operatives to get their requirements. This is pushing temptation too far.

Reference has already been made to the fact that store sheds, etc. must be erected on level ground, but if this is not possible, it bears repetition that the open underneath must be effectively protected. However, as stated previously, it is accepted that you cannot make a store into a fortress.

Keys

The minimum number of key-holders is a necessity and only staff should be so trusted. It is elementary to condemn practices of leaving the office or store key under a stone, at the end of a scaffold tube or hanging on a hidden nail, but in spite of all advice, cajoling and threats such bad practices still operate. The only keys to be taken off site are those for the office block and then only by trusted staff employees.

All other keys for compounds and stores must be kept in a locked cabinet or drawer in the office. They should of course be suitably labelled but not so that any unauthorized person can identify them. A code of lettering or numbering can confuse such persons and the 'key' to such coding must be stored separately.

The same system can apply for keys to partially completed units on the contract being worked, and the practice of having a board on the office wall with all keys, neatly in rows and labelled, is not recommended, unless it is capable of being covered and locked like a very shallow cupboard. Even then the code identification system is advocated.

Specialist storage

It would be ideal if all materials were delivered into and stored in a compound, but the additional costs of transporting from the compound to the scene of operations are often so prohibitive that many companies now arrange for many materials when delivered to be stored outside, adjacent to the place where they will be used.

The following security suggestions for specific articles are recommended to try to reduce losses due to theft and damage.

Timber

Ideally stored in a compound, but even so timbers must never be stored on the ground; a temporary staging of scaffold poles should be used with overall polythene cover for weather protection.

There must be a system of segregation in such temporary staging for the varying types and sizes of timber, e.g. floorboarding, architrave, angle beading, skirting boarding, otherwise extensive damage and wastage can be caused by workmen searching for what they want. Furthermore, in this context, it is not unknown for joiners deliberately to use architrave instead of beading to cover over plastering defects, etc. thus confusing the supply and demand means of assessing requirements.

It is worthy of note that a good site manager insists on all joiners on his job having a nail box, for it is well known that this reduces losses and wastage, and furthermore reduces on-site punctures.

In the context of timber security, site staff, when investigating an alleged shortage of timber, must always take into account that used for office tables, shutters, spot boards, canteen shelves, stools, etc. either made or repaired on site.

Bricks – facing

If loose they must be neatly stacked on wooden battens to avoid damage from the weather generally, but more particularly so they are not chipped.

If they are delivered in pallet or banded form, they should be stored on battens for similar reasons, but the packages must not be stacked on top of each other because when the band is broken, bricks may well be damaged by the resulting fall. The lower the fall the less the damage. Site management must also be constantly vigilant that operatives are not allowed to throw facing bricks into say a dumper bucket for transportation but place them carefully. These elementary precautions can prevent high losses of facing bricks, although it is accepted that where possible it is customary for chipped facings to be used as commons.

Bricks – common

The method of delivery is usually governed by the charges agreed by the buyer and if in pallets or loose, the recommendations applicable to facing bricks equally apply. Sometimes, however, common bricks are delivered by tipper lorry and tipped where required on site, but this is not recommended for although chipping is not as vital with common bricks, the losses by burying can be enormous.

Breeze blocks

Because they are generally more fragile than bricks, on site they should be stacked on edge like facing bricks, and one cannot stress too often the damage that can be caused by 'tipping' or 'piling' breeze blocks.

Window frames

These can be stored on site where required but levelled battens should be used as a base. The frames can be stacked on top of each other up to about six high and while it will not prevent a determined thief, a wooden stake hammered into the ground, and nailed to the top window frame will hold the others in position and prevent damage by vandals. Some builders may be horrified at this suggestion but they cannot refute the claim that if a thief goes to a site for a window frame the obstruction of a compound wire is not going to deter him.

Glass

This should be very carefully stored in a hut, ideally 3° out of vertical on a wooden base, on its own if possible, but if not, sufficient space must be allowed for men to pass it with ease. Cumbersome articles, e.g. sink tops, etc. must not be stored so that removal necessitates passing the stacked glass.

Plumbing and electrical goods

These are some of the most valuable materials on site, and because of this it is essential to use the best stores for them. Usually the corrugated metal type is used, and if so, it is more important, as stated previously, to ensure that all fixing nuts are on the inside, that the roof is securely bolted down, and as this type of store generally has no floor, that access cannot be gained by excavating under the walls. It is advisable to fix small-mesh covers over the windows in this type of store, fixings on the inside, but just in case the mesh is forced off, ensure that no small items, such as copper fittings, electric plugs and switches, etc. are placed so near any potential opening that they can be lifted out without entering the store.

In some thefts from building sites, particularly when a large quantity of plumbing or electrical materials is alleged to be missing, it is pertinent to observe that many times management must have been very unwise to have such a large stock on site as it could cover weeks or possibly months of work ahead.

It is a golden rule that only sufficient stock of requirements for a short period should be held on a site and this particularly applies to expensive plumbing and electrical gear.

Cement

Bags of cement must be stored under cover in a hut with a floor, wooden if possible, and on a sheet of polythene so as to prevent setting due to dampness. Ideally such a store should have a door at each end so that the cement can be used in the correct rotation, and special efforts must be made to make the roof waterproof. It is not unusual to find the bottom two layers of bags solid because of neglecting these factors, and this is a straightforward avoidable loss if thought is applied. It is often said that such solid bags can be used as 'back fill', but because of the large solid shape this can cause many later problems.

Concrete paving slabs

These are not usually delivered until near the completion date of the project and as such can be stored out on site where they are to be laid. The maxim that a determined thief is difficult to combat applies very much in this case, but the costs of moving these slabs about the site could be quite prohibitive for, because of bulk and weight alone, the use of transport is a necessity.

Flimsy materials

Many materials are very damage prone; for example, kitchen units until fitted are very fragile as also are plaster boards, hardboards, metal ventilation ducting and suspended ceiling fixtures. Obviously greater space must be provided for the storage of these items and if of varying sizes such as the kitchen units, the sizes must be clearly indicated to avoid confusion when searching as this causes damage.

The case is recalled of an expensive, ornately carved font for installation in a church being built, and provides a good example of specialized storage, where the slightest chip or damage would have been catastrophic. Due to circumstances outside the control of the builder, the font was delivered some weeks before it could be installed, and the site in a populated district was 'tight'. However, because of its fragility, it was decided to restrict any further transport and to store it on site. To ensure complete protection, it was off-loaded into the corner of the compound, a brick wall was built around it and topped off with secured, bedded concrete paving slabs. The expense was justified for the font remained intact.

In all cases of flimsy materials, some indication as to the ideal storage can be gleaned by noting how the supplier delivered it, i.e. vertical, flat, etc. and trying to do likewise.

Pre-formed roof trusses

These should be stored vertically as they are usually delivered by the suppliers, and should be placed on wooden battens to avoid damage. Alternatively, they can be stored upside down on a firm scaffolding frame, ensuring that the apex is well off the ground.

Door casing, staircases

When stored, but particularly after fitting, they must be protected to avoid damage by passing workmen when using wheelbarrows

31

on the former, and unguarded treads on the latter. It is not unknown for such damage to warrant a complete replacement.

Reinforcing wire

Should be stored flat and as it should always be labelled when delivered on site, the label must remain intact until the whole of it is used to avoid wrong use, particularly if various types and sizes are being used. Labourers are prone to use from the nearest stack.

Tiles

Roofing tiles should be stored on edge on level ground and covered with a sheet or polythene to avoid splashing.

Hanging tiles for outside use are usually very brittle and careful storage is required.

Glazed tiles for internal use are best kept in the boxes provided, in a store shed until they are fixed.

It is essential that only the proper cutting tools are used on any type of tile.

Oil drums

If the fuel oil is stored or drawn from drums on site, it is advisable to have them on temporary metal stands made from short pieces of scaffold tube and to have an adequate tap and a suitable container (measure) to restrict wastage when drawing a supply.

General

It is impossible to cover the multifarious types of materials required and used, and therefore stored on building projects, but the question of how and where stored, whether or not in a compound, is affected by many factors – the risk involved, the costs of compounds, the cost of loading and transporting from compounds and stores to the scene of the operation, the possibility of damage during storage and transportation, and the easy access to stores, etc. due to the progress of the project. It behoves every site manager to ponder these factors at the beginning of a job, for a little thought can often save subsequent expense and frustration.

One fact which cannot be disputed by anybody is that neat and tidy stacking of all materials takes up less room, usually at a premium on a building site, and if there is any violation from theft or damage it is immediately obvious.

Sub-contractors' materials

It is absolutely essential at the outset to know who is responsible for the storage of the materials of sub-contractors, whether nominated or otherwise. If this is not clear it is the duty of site management to get to know, otherwise costly complications can arise as to responsibility for damage, etc.

7

Out of hours security

The first measure to ensure adequate security when the site is not open is to check that all normal precautions have been attended to, viz. all doors locked, shutters secured, compounds closed and locked and all loose tools, e.g. transformers, drills, angle grinders, lump hammers, picks, shovels, saws, etc. collected from site and locked up and partially completed units secured. All machinery left on site such as mixers (drum cleaned out and inverted) and dumpers (in compound if there is one) must be immobilized; ladders should not be left erected on scaffolding or, if they are, a batten should be roped over the staves to prevent climbing by trespassers as far as possible. Any easily transportable valuable materials, e.g. unfixed metal clothes posts or gates, should not be left loose on site; the list is endless, but sound common sense should prevail and it is a site management responsibility. There is an erroneous view by some site managers that if a watchman or a security firm is engaged that responsibility is transferred. *This is not so.*

Even though all the precautions listed have been taken there must be some type of security when management is not there and the following details some constructive suggestions for cover.

Watchman

There are many ways of providing watching cover during non-working hours, the most common being either a full-time watchman, i.e. from closing-down to opening-up, or a part-time one with fixed hours, or casual visits by a watchman or watchmen.

Much depends on the locality and the length of cover is governed by this. The type of man engaged is also vital, for it is generally accepted in these current times that if he is conscientious enough he is usually too old to chase off trespassers and vandals, and if young enough for the latter, the wages as laid down for full-time cover under the National Working Rule 2.8 are not sufficient to attract him. The vital aspect is that work from site closing, say 5 p.m. to site opening, say 8 a.m. the following morning, counts one shift and the pay is at labourer's rate plus guaranteed minimum bonuses.

However, if a watchman is engaged, he must have access to the general office telephone to call for police or other assistance and he must be clearly instructed on his duties. Because he must be aware of the location of various stores he must be trustworthy and of good character for he is in sole charge when the site is closed. Therefore all potential watchmen must be vetted, possibly by discreet reference enquiries. The local police, if contacted, may be able to suggest suitable persons for watching duties.

Own employees

Many firms supply caravans for occupation by staff on site and one can argue that this is the best type of security. They know the position of everything on site, and are usually loyal and conscious of their employers' goodwill. Some recompense is usually agreed for employees covering this aspect and such a system works well and is recommended.

Sometimes an employee who lives locally can be utilized as a watchman, and usually this system works satisfactorily. Again, however, he must be trustworthy, and the mere fact that he is a good labourer does not always justify trust in him as a watchman. Such a man can either be for fixed hours or casual visits, but it is difficult to check on him if the latter system is working. He must not be one who is likely to clear off the site to a hostelry, for local thieves will quickly realize the times he is likely to be absent, and possibly by 'treating' him they may ensure that he is away long enough for their operations.

Security firms

They can be employed on site in similar ways, full time in uniform, with or without a dog, referred to as a static guard, or casual visits, but such security cover is liable to be comparatively expensive. However, if the proper enquiries have been made prior to tender stage and, because of conditions it was considered necessary, then an amount should have been included in the tender figure.

If a static guard is used then he must also have access to an office

35

with a telephone for emergencies and it could be used by the site manager (at home) for checking calls to ascertain that the guard is still on site. Instances are many where watchmen have used such telephones extensively for their own private purposes, and consideration should be given to having a 'Pay phone' – a subject dealt with more fully in Chapter 9.

If casual arrangements are made, the security company will agree to visit the site a certain number of times in a given period, and that the times of these visits will be varied for security reasons, but a record of these times will be kept and subject to inspection by the client at any time.

Similarly, the security company will agree to install 'clocking' points on the site whether it be a static or casual system, and that all such clocking tapes will be available for scrutiny by the client.

If such systems are used it is a very bad policy for the client, i.e. the builder, to fail to make periodical checks, for it keeps the security company on its toes and will never be resented by a reputable security company.

Although security companies are apparently expensive never be tempted to accept a cheap tender for security cover. One has only to realize that, as the law stands now, a convicted thief, just out of jail, can have some visiting cards printed and set himself up as a security contractor. Efforts by the International Professional Security Association to get legislation to combat this anomaly have failed to date.

A golden maxim when considering security cover is that the cheapest is not always the best, and it is futile for site management to lay out monies on site-security measures for them to be defeated by a lack of out of hours cover.

It is very important to observe at this stage that a site manager must never disclose or discuss what security measures are operative with other than his supervisors. Very often representatives of various suppliers visit sites and regularly ask the question: 'Who supplies your ..., because my company, who supplies ABC Ltd, would like to give you a quote?' In most cases such facts are openly discussed with the usual reference to the office buyer. However, if an alleged representative calls on site, produces a visiting card ... Security Ltd and asks a similar question, details of existing security must never be disclosed or discussed. There are many cases on record of thieves making this initial approach for it may save them hours of observation over the site.

Site managers must always be wary of two persons visiting the site, for while one might engage the manager in conversation, the other may wander off to 'case' the site.

It is worthy of reiteration that if a recognized security firm is engaged on a written contract for site security, even though they can be regarded to some degree as sub-contractors, they cannot be held responsible for any thefts or vandalism which occur during their cover, even though they are insured. Many actions have been taken in the courts for this, but all have failed because negligence could not be proved.

Police

The fuller aspects of liaison with the police are referred to in other chapters, but while you will never expect nor get the police to patrol a building site, unless there has been a spate of thieving or vandalism, they will, in the course of normal patrol, visit the site, if they are so requested.

It is an admirable ploy to encourage a policeman, in uniform, to visit the site in working hours. The local constable, whether he be called community officer or other designation, can be of inestimable value as a crime deterrent for his mere presence, like that of a police patrol car on a motorway, has a salutary effect. Furthermore, he can see who is employed, possibly for future reference, and will certainly deter any potential thief 'overlooking' the site while the policeman is there.

Alarms

There are many types of burglar alarms available for use on a building site. They are all electric circuits, either mains or battery, on doors, windows or pressure mats, and when broken they either operate through the site telephone on the '999' system or direct to a security control and/or trigger off a loud bell or klaxon or flashing lights. With the telephone system it is hoped that the thieves will be caught and, while the police encourage the installation of alarms, there is such a high percentage of false alarms (well over 90%) mainly due to wrong activation, called subscriber error, that rightly the police often give notice that they cannot always guarantee a response to particularly troublesome systems. However, it is hoped that the bell ringing will frighten off any intruders and if so, the alarm has justified its installation.

Alarm systems, some of which are portable and easily installed in new locations, can either be bought or rented and maintained by reputable suppliers, but the main cause of failure is inefficient handling when being put into the operative position.

When exploring the possible use of a burglar alarm, it is always advisable to get at least three quotations for the required needs so as to get the most economic system.

One could go to great lengths describing the various types of burglar alarms, but suffice it to record that vibratory alarms, i e. where the act of knocking a door as opposed to forcing it will operate the alarm, are useless on building site offices for they can easily be operated by the wind. The beam type, i.e. where the alarm works if a beam is broken, is ideal for the inside of a store building holding valuable goods, e.g. plumbing or electrical goods, but is not very practicable when used outside. Often beams are double banked to catch the crawler and the walker but outside on building sites there have been cases where animals have activated them and also where dense fog which arrived during the night time triggered off the alarm.

There are two schools of thought on the publicity to be given to the fact that a burglar alarm system is installed. One, that it should be kept secret, but the other advocates publicity, for it is a well-known fact that thieves are not attracted to alarmed premises, often evidenced by prominent display of the alarm bell in an almost inaccessible position.

However, if publicity is given it must be of a general nature and should not, in any circumstances, give details of the type of installation or which stores, etc. are covered.

Sophisticated burglar alarms can be expensive but many security firms now sell the necessary parts for a system in a DIY pack, with fixing instructions well within the scope of an average electrician. Doors, windows, roof lights, etc. can be guarded, but it must be realized that these comparatively inexpensive systems only activate an alarm bell and do not make any contact with the police.

The author had used these 'home-made' alarm systems with success for many years before the advent of the DIY packs in their present state, but it must be stressed that it is essential that care in dismantling is just as necessary as when being installed if they are to be used over and over again with satisfaction.

The small, portable type of alarm, which can be hired for short or long periods, is recommended for temporary cover of particularly vulnerable property, e.g. show houses or completed buildings ready for hand-over.

If there has been a spate of breaking into site offices, etc. and the facts have been reported to the police, they will often lay or fix one of their own pressure mats, which if activated 'bleeps' to the local station. Obviously if such an installation has been made it must be kept as secret as possible and no doubt the police will arrange for the fixing of the mat during non-working hours.

The decision as to whether to fit a burglar alarm to any particular store is the responsibility of site management in consultation with any staff security service the company may have, who must compare the necessary expense against possible losses having regard to the type of materials protected, the strength of the store, the location of the store on the site and the locality of the site. In common with all security measures, it is impossible to estimate the beneficial effects of an alarm system, but with very high premiums the insurance company could well offer a reduced rate for property covered by an alarm system.

Floodlighting

This is an admirable deterrent, for it is widely accepted that thieves do not like working in the light. Again there are two schools of thought on how they should be fixed. If a compound is to be flood-lit some advocate high poles on the compound perimeter, with the lights facing inwards, while others install a central pillar with the lights facing outwards. If the poles are on the outside perimeter then the poles supporting the lights must be adequately protected with barbed wire against climbers.

The type used depends very much on local conditions, such as causing annoyance to nearby residents during the dark hours.

There are, however, two cardinal factors which must be observed if floodlighting is installed:

1. The glasses of the lamps should have a wire shield to prevent damage by stones, etc. It is regretted that little precaution can be taken against the use of an airgun or the like, apart from fitting armour-plated glass on the lamps, and
2. An automatic time switch is an absolute necessity, for not to fit one is to depend on human frailty, particularly if the floodlights are to be activated some time after the site has closed.

One must again stress the importance of the careful dismantling of a floodlight system for subsequent use on another site, otherwise it can be an expensive operation.

The Electricity Council publish a free booklet entitled *Essentials of Security Lighting*, obtainable through any Electricity Board Office.

Guard dogs

There is no doubt that dogs are a very valuable asset to watchmen on building sites in preventing thefts or damage and arresting offenders, but since the well-known case of a child crawler who got under the compound wire and was savaged by a guard dog, the indiscriminate use of dogs is no longer lawful.

This incident was responsible for the hurried passage through Parliament of the Guard Dogs Act, 1975 which clearly states that guard dogs are not allowed at any premises unless under the control of a handler at all times or so secured that they are not at liberty to go freely about the premises.

Translated into the practical use of guard dogs on building sites this means:

1. A watchman can use a dog, provided it is under his control. Whether this means on a lead or only subject to immediate obedience to words of command is not clear.
2. Clearly the formerly accepted practice of leaving a dog without a handler to roam at will in a compound is unlawful, but it would seem to be allowable to have a dog alone in a compound if it is tethered, possibly on a fixed lead attached by a running loop to a taut wire. This would allow the dog to traverse the compound area, tethered as the law states, but not so free as to enable it to get to the compound wire perimeter. A gap of 1 metre (3 ft) is recommended between the limit of the tether and the compound perimeter. All such tethered dogs must be visited periodically to ensure that the dog has not injured itself or been injured by intruders or got loose.
3. Similarly, an unattended, loose dog in a suite of connecting offices or a particularly vulnerable locked store is not allowed, but it can be argued that a tethered dog in such circumstances would have the desired results if barking alone puts off any potential thief or vandal.
4. In all cases where dogs are used, suitable informative

notices must be prominently displayed. Again it can be observed that these notices alone can have a marked effect as highway notices such as 'Speed Check Ahead' have on motorists.

5. It is fair to observe that since this Act was introduced, prosecutions have been very rare; in fact so far as is known the question of court action is only ever considered if some trespasser/offender has been severely savaged.

8

Safes, cash, wages, etc.

Safes

It is generally accepted that cash ought never to be left on a building site, but much depends on the type of site, the length of the project, etc.

Some companies, however, allow the practice of leaving money on sites and a natural consequence of this is the necessity to provide greater security safeguards, which may range from the use of a metal filing cabinet type, to a proper safe bricked or concreted in, depending on the value being kept.

It is the firm view of many experienced security advisers that the mere fact that there is a safe on a site presupposes that there is much of value to protect, namely cash, and thieves may be attracted. This supposition opens up a whole new aspect of crime prevention, for thieves who tackle safes are generally of a more determined type and difficult to counteract. It is common sense not to attract safe-breakers and safe-blowers on to building sites, so in general terms safes are not recommended.

Many advocates of having safes on sites argue that in addition to cash, the documents of new employees recruited on site, e.g. Forms P45, holiday stamp cards, delivery notes and other vital documents must be safeguarded, but these documents are rarely permanently on site but should be sent as soon as possible to the head office where they are less vulnerable.

If a safe is provided on a site the key for it must never be left on site when unattended and obviously the 'storing' of the key, suitably labelled, in a drawer of the agent's desk is ridiculous.

Cash

If safes are not used on sites the question is raised as to where to store site cash safely, which is answered by asking the question, 'Which site cash?'

Excluding site wages which will be dealt with later in this Chapter, usually the only cash held on site by management is the site 'float' which is run like an imprest account. It is accepted that a manager needs ready cash for small purchases for use on site such as domestic articles or small materials urgently required, the absence of which would delay the job if it were requisitioned in the usual way, e.g. new hacksaw blades, etc. but otherwise all other aspects, such as cash for pay-offs and subbing to employees, are sent from head office when requested. Thus a manager is given a site float at the outset, an amount being dependent on the size of the site and anticipated expense. He makes his purchases and sends all his vouchers to head office at regular, possibly weekly, intervals, when he is reimbursed to bring his float back to the original amount. The site manager should carry the float on his person avoiding the use of safes, etc. and if he is efficient he will be sure that he is not in possession of too large a float, which in any case should be carefully monitored by the head office and only sufficient to cover anticipated expenses.

Subbing

Mention must be made of the 'subbing' system prevalent in the building industry. This is really an advance on wages due and site management must always be certain before a decision is made to 'sub' that the applicant is already due that amount of wages. If an advance is made it is of paramount importance to obtain a receipt signature against the advance, such receipt being sent to the head office for adjustment of the employee's next wages. It is good policy to record the amount in figures and words, e.g. £10.00 (TEN POUNDS) on any receipts.

Efforts made by companies in the past to prohibit subbing have failed and consequently many impose restrictions on the amount advanced and the period, subject that at all times the applicant must have already earned that amount; some examples follow below.

Directly employed labour

First week – 75 per cent of entitlement up to a maximum of £30.00. Second week – 50 per cent of entitlement up to a maximum of £20.00.

43

Third week – 25 per cent of entitlement up to a maximum of £10.00.

Labour-only employees (sub-contractors)

First week – £30.00 per man maximum to be paid by cheque. No subbing allowed after one week's employment.

Site management must never lend or advance money to an employee without an official record being made, for cases often occur where unofficial loans made to employees are denied, to the embarrassment of the site manager.

Wages

Ideally all wages, bonuses, expenses, etc. should be paid by cheque, but for reasons which cannot always be understood, attempts to standardize such a system in the building industry have not been successful. Consequently the various ways of paying wages on sites must be explored, and their advantages or otherwise must be highlighted.

1. Wages made up at head office and transported by company transport direct to site. Security demands that the wages should arrive after the lunch break on Thursdays, the nominated pay-day, and therefore the wages are on site a minimum of time before pay-out. If a company has many sites this later delivery time is not always possible, therefore secure storage is necessary on site. This may suggest having a safe, but if the wages are stored in a metal filing cabinet in an office which is not left unattended this would appear to be adequate.

   Invariably this system uses two men from head office for the transport, or if the site staff collect wages from head office two men are employed.
2. Similar procedure, but instead of using company men for transport, the services of a security firm are utilized. This could prove expensive if the amount of wages does not justify it.
3. Cheque with breakdown of cash required and wages sheet sent to the site manager who, with an assistant, goes to the local bank, obtains the money, then makes up the wages in the site office. There is an obvious security risk on the

bank journey and when making up the wages and two site men, particularly the site manager, are being used for non-building work.

4. Wages made up at head office and sent by registered post to post office near the site, to be collected by site manager, authorized in writing to collect. This system is good for small sites, but there is a security risk on the journey to collect from the post office, and complications can arise if the post deliveries are late. As head office staff are used to post the wages, where two men are needed this is a man-power waste and there is also a security risk on this jour-ney, particularly if there are many site wages to post.

5. Wages made up at head office and transported by security firms for registered posting. This can be expensive.

6. Pay slips posted direct to site, cheque and wages totals breakdown posted to security firm, who cash the cheque, make up the wage packets and deliver them to the site. This is particularly good for sites well away from the head office but of course can be expensive. However, one has to assess how many firms' man-hours are saved by this method and compare it with the costs of using this system.

   The costings of all 'cash-in-transit' schemes using secur-ity firms are negotiable, but if this system is adopted, the observations about 'cowboy' security firms equally apply, for they may not be fully insured for this work and the firm's own insurance rarely covers cash handling by out-siders.

7. The labour-only type of sub-contractor who has been granted a Form 714 Certificate (exemption from the stop-page of income tax, etc.) should always be paid by cheque. Experience reveals that they are not enamoured of such a scheme and would rather receive cash so that the share-out can be made there and then on site, but responsible build-ers are adamant that if such sub-contractors have qualified for the exemptions shown, then they should be sufficiently business-like to have a bank account.

Paying-out

The method of paying out wages varies between companies. On small sites the site manager delivers it personally, while others have a full pay-out at an appropriate office. In both cases there are security risks and there are cases on record of raids by thieves at pay-out points, whereas the personal delivery rarely attracts wrongdoers. It may be that the size of the pay-out governs the temptations. If there is a central pay-out, usually on large sites, the office must be substantial, well locked up and secured with only the pay-out window open. There must always be a telephone in the pay-out office, and facilities for sounding a bell or other alarm for assistance.

If a bell is to be used to signify an emergency and a call for assistance, then it is absolutely essential for everyone on site to know that such a bell is in existence and likely to be used. Test alarms at pre-arranged times must be made to check the continuing efficiency of the bell and to familiarize the site employees with the noise so that, should they hear it, they know that a 'raid' is under way.

In many walks of business and commerce, when the anticipated method of payment is by cash, a signature for receipt of the wages is usually obtained but for some unknown reason signatures for wages, etc. on building sites are practically unknown. Because of this it is absolutely essential that the person paying out knows the employees so that mistakes are not made.

Reference must be made to cases where one employee requests another man's wages, possibly due to sickness, and it is paramount for that requester to be known to the paymaster and for a receipt with signature to be obtained.

Great care must be taken to establish identity when a person, other than an employee, calls at the pay-out to collect the wages of an absent employee, or one who has previously left the site, for it is not unknown for thieves to try this method, or for a disgruntled wife or friend to attempt to get hold of a man's wages.

Any wages, not collected, should not be held on site but returned to head office. This may seem irksome if the man calls the next day for his wages, but it should be explained to him that it is laid down that pay-day is Thursday and wages cannot be held on site pending his ability to attend.

9

Miscellaneous matters and measures

Staff purchases

Nearly all companies allow their employees the privilege of purchasing spare materials from sites, but this has to be very carefully monitored and recorded. In all cases there must be a written record of the transaction and the employee must be given a written authority, specifically naming the article(s) to accompany the transport away. The police know of these widespread arrangements and if anyone, employee or otherwise, is seen taking property from a site he will be arrested unless he can produce written proof that the transportation is lawful. The site manager must sign that authority and he must report the fact to head office, who will agree the price. There should never be any money handed over on site without head office approval.

Off-cuts

Unless specifically authorized in writing, site managers should have no authority to give away anything from the site.

It is a fallacy, although some ill-advised companies allow it, for tradesmen to be allowed to keep off-cuts of electric cable, copper tubing, etc. as a 'perk', for it could encourage employees to make extra off-cuts. There are cases on record of electricians deliberately cutting cable into short pieces to be burnt to get at the copper.

In some companies all scrap metal is returned to the office for disposal, but in many companies power is vested in the site manager to dispose of any such items by sale and all dockets covering the sale are forwarded to the office with a cash adjustment in the site float. Experienced supervisors are quite capable of checking the expected amount of scrap against the amount recovered.

The taking of short off-cuts of timber for firewood could

47

encourage longer off-cuts, and the best policy, regretfully in some ways, is for all scrap timber to be burned on site.

A prosecution for stealing already fixed setting-out pegs failed because the thief alleged that he thought the wood he had taken was scrap as the foreman had previously given him similar pieces of wood.

Site management must be very firm and determined on these aspects. Trespassers must not be allowed on the site in any circumstances for they may be tempted to take something they think has been thrown away.

Found property

Often during excavations, generally on demolition sites, old electric cables valuable for the copper content, old lead pipes and similar materials are discovered having been buried during previous operations by other contractors. Some workmen are of the opinion that 'finders are keepers'. Such is not the case, for the law is very clear that all such property found belongs to the owner of the land, unless there is a reference made in the contract, when it becomes the property of the contractor. It cannot be stressed too much that such found property never belongs to the workman who finds it, and if he takes it, it is theft.

Car parks

Indiscriminate parking of cars by employees must never be permitted on site, and arrangements should always be made for a central car parking area. The wisdom of this is abundantly obvious, for one must accept that if a car is parked in an isolated part of a site it is much easier to put something in the car boot, e.g. a sink or worktop, than to carry it openly across the site to be put in the boot.

All employees must be informed, and a notice must be displayed, that all cars left in the official car park are left at the owner's risk in that the company will not accept any liability for damage caused by other employees' vehicles. If, however, a car is accidentally damaged in the car park by company transport, e.g. a dumper, then the company could be liable. If a car is parked anywhere other than the designated parking area, the company should not accept any responsibility under any circumstances.

Fire precautions

It is well known that the temporary accommodation for offices, stores, etc. erected on building sites is generally of wood, and particularly as a means of heating is either propane gas, electric or natural gas, fire is a constant hazard. A fire on a site can have a disastrous effect if drawings, records, variation orders, etc. and of course materials are lost. It therefore behoves every site manager to ensure adequate precautions against fires and as far as possible that he is prepared for such an emergency. Action is much more urgent than for normal fires, for the usual huts and stores are easily and quickly destroyed. Therefore some fire precautions are necessary at the outset.

Listed are some of the main causes of fires on building sites.

1. Allowing clothes, particularly when drying off, to hang over either open gas fires or the convector type. Hanging hooks must never be allowed to be fixed or remain in such positions.
2. Siting fires adjacent to wooden doors or furniture, for it does not take very long for the wood to dry and start to smoulder.
3. Unattended canteen water boilers left to boil dry. It is customary for one man to be detailed for this supervision, but this is negatived if all and sundry are allowed to draw hot water without refilling.
4. Unqualified electricians carelessly wiring up plugs, light sockets, etc.
5. Lighted cigarettes left unattended in ashtrays and very often lodged on the edge of wooden furniture.
6. Storage of inflammable articles near to open fires.
7. Bad connections to propane gas bottles, possibly caused by using rubber tubes, pushed on to the connecting joints, becoming bell shaped and loose. Only the special type of tube plus jubilee-type clips should be used for connections. It is worthy of note that gas bottles must be at least 2.5 metres (8 ft) from any open fire because of the obvious risk.
8. Bad storage of propane bottles and rough handling, for example dropping them on the ground from the delivery wagon, doing damage to the fabric not noticed at the time, but the cause of later leaks.

Propane gas bottles when not in use should either be stored in the open, with a substantial chain threaded through the handles and locked for security or in a small wire compound, such as used on the railways or, if in a hut, then the hut must be suitably labelled so that the fire service, if called to a fire, are alerted to the danger.

9. Bad siting of petrol or diesel tanks. Obviously these must be away from wooden huts and should be securely locked when not in use. All tanks are fitted with taps, but such is not always the case when the liquid is stored on site in 200 litre (45 gallon) drums. Often the practice of tipping the drum to fill a bucket, with consequent spillage (wastage) and leakage can constitute a fire and security hazard. An adequate funnel must be used with this method and also when dumpers, or other machines, are being refuelled by the bucket system.

One could outline many causes of fire which can very often be prevented by sheer common sense, such as prohibiting smoking in timber stores, or clear thinking on fire prevention. NB: *The fire station telephone number must be sited very clearly above the site telephone.*

One must re-emphasize the swiftness of fires on building sites and even though the fire service quickly answer the call, it is usual for irreparable damage to have already been caused.

Because of this, it is essential to have fire-fighting apparatus easily available and while the fire service will readily offer advice on the type and number of extinguishers which should be on any site and office complex, one can always use buckets of commodities which are readily available on building sites, namely sand and water. It should be stressed that these buckets when filled are not ashtrays and they should be either hung on the wall 0.3 metre (1 ft) above ground level or stood on bricks two high. The number of fire buckets, which should be suitably labelled, depends on the size and number of offices, but one of each should suffice for a small wooden building for it must again be stressed that immediate action is called for if a fire breaks out.

Fire precautions in partly or completed structures are again common sense for if a particular operation is likely to cause a fire, for example use of a salamander type of drying-out machine, then fire precautions are essential and must be readily available.

If sophisticated fire extinguishers are located in offices and on sites, the site manager must ensure that his staff likely to be involved are fully conversant with the means of operation and that the extinguishers are serviced by experts at the recommended times.

Clocking-in offences

If there is a system of signing-in or clocking-in on a site at the start and finish of the work period, it must be made very clear by notices posted at the clocking signing-on position, or in a note in each man's pay-packet that in no circumstances will the signing/clocking-in of another person be permitted. Failure to comply should warrant instant dismissal. Management must be very firm on this score, for such actions are nearly always a prelude to a 'fiddle'. All sorts of excuses are advanced when caught, such as 'I thought he was on his way', etc. but rarely does it arise that management are told that a mistake has been made in time to prevent extra, unlawful wages being paid.

It is appreciated that a mistake can be made but experience shows that such cases are very few indeed.

Any workman found fiddling with the clocking system should be dismissed immediately and reported to the police with a view to a prosecution.

Overtime

As a general rule, overtime should never be permitted unless it is adequately supervised for it is a security risk. Questions as to who locks up the stores, turns off all power supplies, including calor gas boiler, etc. always arise, and if unfortunately something goes wrong, the standard excuse of 'I thought he had seen to that' is not a satisfactory answer to a major theft, damage or a fire.

Trades' rubbish

It is the responsibility of all trades to clean up their own rubbish after their operations, but many try to get away with not doing it. It is understandable if they are on bonus targets, but site managers must be firm in insisting on the 'clearing up of own rubbish' system. Tradesmen should be told in no uncertain manner that any charges incurred in clearing up after their work will be deducted from their dues. Unless a tight rein is kept on this matter, the company can be involved in considerable expense.

Opportunity is taken to stress the importance of checking rubbish skips and containers going off site for secreted stolen property.

Transfer notes

The procedure on receipt of materials that a signature is only given for that actually received, is equally important when materials are sent from the site, either returned to the supplier, transferred to another site or the head office. In all such cases details should be recorded, and many companies have triplicated forms on which the signature is obtained. The site keeps one copy, two copies travel with the goods, one of which is signed and sent to the office for the buyer's and accountant's attention.

Telephones

On most sites it is essential to have a telephone and the use of that telephone should be restricted to the obvious reasons. Personal calls should not be allowed, neither should sub-contractors be allowed to take advantage of it. Many building firms, in a spirit of full cooperation, have permitted such uses to their cost. Cases are on record of sub-contractors' men occupying the telephone for over an hour in peak time giving details of the hours worked by them to their head office. Managers must be very firm on restricting such use of the telephone. It is possible to obtain very cheaply and fit a small locking device on the telephone dial to prevent unauthorized use.

Many firms, frustrated by persons taking advantage of the site telephone, now arrange for a pre-payment phone to be installed to monitor the use. In theory the company should benefit from such a system for call-box rates are made for each call but the company is charged at the normal rate. If a pay phone is operated then further scrutiny and thought must be given to the emptying of the coin-box. Some companies allow the site manager to empty the box and adjust his weekly site float return with the proceeds, but others do not allow the box key on the site, it being kept and used by the visiting supervisor or surveyor, who checks the contents with the site agent to prevent any suggestion of fiddling.

If the wages pay-out is from a particular office, and this is only usually on big sites, then it is essential that a telephone be easily accessible for emergency use and similarly a night-watchman should have access to a telephone.

Foreman's diary

Every site manager must keep a fairly comprehensive diary of happenings on the site. References to names of visitors, type of weather (for wet-time payments), details of verbal cautions for disciplinary purposes, and any other happenings, such as discovery of a theft, which are very often vital in the field of security.

Electric and gas charges

On every type of new buildings where these services are to be utilized, application is always made for the installation by the builder, and because of the accounting system his name is recorded as being responsible for all payments including the energy used. The installations must be tested for efficiency and safety, during which time certain energy is used.

It is accepted that sometimes gas and electricity are used for drying out, lighting for end trades, etc. but what is very often neglected by forgetfulness is to ensure that at the time of handing over the keys to the client, the meters are read so that the company's liability for payment ceases at that time.

Powers of search and arrest

Many reward notices on building sites refer to the possibility of persons and vehicles being searched by the management, but it must be emphasized that no such power to search exists in law. However, if a site agent has good grounds for knowing that stolen property is either on a person or his vehicle, he can request permission to search, and if it is refused, he can make a citizen's arrest, which has the full backing of the law, and notify the police and hold the suspect till they arrive.

The legal power of arrest of a private individual is given under Section 2(2) of the Criminal Law Act, 1967:

> 'Any person may arrest without warrant anyone who is, or whom he with reasonable cause suspects to be , in the act of committing an arrestable offence',

and Section 2(3):

> 'Where an arrestable offence has been committed any

person may arrest without warrant anyone who is, or whom he with reasonable cause suspects to be guilty of that arrestable offence.'

These extracts are self-explanatory and in this context it is only necessary to appreciate that the words 'arrestable offence' include thefts and cases of damage.

Temporary order books

When the requirement is too great or too expensive for a cash purchase, many suppliers will not accept verbal orders, but require a written order with an order number. This has lead to the introduction in some companies of supplying sites with a written order book, described by various names but usually called a temporary order book. The order issued from this book by the site is later confirmed by a regular written order from the buyer. The use of these temporary order books is not recommended, but if used, they must be carefully monitored in that only named persons are allowed to issue such orders, for all debts incurred have to be paid by the company.

10

Damage

Reference has been made elsewhere to the percentages of acceptable losses of some materials in the construction industry, e.g. 5 per cent for facing bricks, but in practice these allowances are greatly exceeded to the financial detriment of the builder. Much of this loss is affected by the efficiency of site foremen to supervise correctly tradesmen under their command. There are certain codes of practice governing the work of these tradesmen but, because of the bonus target system, side-tracking is common.

A few examples will illustrate the wastage, which in this context is another name for damage, which is readily visible on every building site.

1. Broken bricks. For the sake of quickness, a trowel is used instead of the proper tool to halve a brick. The brick does not always break evenly and both parts are then thrown away. One only has to inspect the bottom of any scaffold to see the evidence.

2. Mortar droppings. It can cost considerable expense, at the maintenance period, to rectify these either in the cavities, on wall ties or on fixed window sills, whether wood, brick or stone.

3. Timber off-cuts. An inspection of any building site will reveal that the amount of wastage by large off-cuts of floorboarding, architrave, skirting boards, quadrant, etc. thrown around the site could and should be curtailed by strict site supervision to ensure that the tradesmen perform their work efficiently and not just speedily.

4. Damaged doors. While it is accepted that some doors may be damaged in the store sheds, much damage is done out on site. Take the flush paramount type when the only solid piece of wood is the site of the door fastenings. Very often, again in the urgency for speed, when the recess for the lock is being made it is not unusual for the bit to come through the side, damaging the door beyond repair. The damaged door is then disposed of by fire or other means to cover up the bad workmanship. The way of checking on this aspect of damage is for a specific workman to be issued only with the quantity for a specific area of work. Any deficiencies then arouse suspicion.

Many will query the extent of this type of damage, but experienced staff know full well that if the correct number of doors are ordered and checked on receipt, any subsequent shortage must either be damage on site or theft. Often theft can be ruled out, so damage is the only answer. Incidentally, it is useless reporting to the police that a large number of doors are still required to complete the site and therefore they must have been stolen. The police are well aware of this damage shrinkage, which can be prevented by better site supervision. It is accepted that no tradesman would deliberately damage a door as outlined, but persistent transgressions must cast doubts on his ability and raise the question of disciplinary warnings, etc. if directly employed or of a claim for damages against him if he is a sub-contractor.

A large amount of damage to materials is done on site by workmen either by carelessness or obvious neglect. It cannot be stressed too often that a tidy site is generally an efficient site, and every site manager must try to educate his men to ensure that all materials are safely and correctly stored, for in this way much damage can be avoided. It is bad practice to stack materials at the side of routes used by dumpers, delivery vehicles, etc. for inevitably at some time damage will ensue. Materials which are affected by adverse weather conditions must be sufficiently protected, if not in stores, then with covers firmly fastened to counteract wind and rain.

Earlier reference has been made to using garages, bungalows, etc. as good stores, but if, for example, kitchen cupboards and units are stored there it is equally important to store them in known sizes, for until fixed, these articles are very flimsy and easily damaged. If a workman has to clamber over or push past to find the size he wants much damage can be caused.

One could give many examples of bad storage resulting automatically in irreparable damage, but the best way to prevent it is to ensure that the site is tidy and neat.

The major type of damage committed by trespassers is broken

windows and the prevention of it creates a serious problem for all builders for, in addition to having to replace the glass, the damaging effect of the weather can be enormous. Watchmen can attempt to keep trespassers off the site, but windows on the perimeter of a site, adjacent to public access, are particularly vulnerable. Many builders have recourse to boarding up ground-floor windows, prior to painting, but this is expensive and can delay the project. Others have fixed wire mesh temporarily over the windows which at least reduces damage by stone throwers, but a more difficult problem is created by offenders using airguns and the like. Window breaking on an almost completed building has a demoralizing effect on the workmen. Such a case occurred at a church, when on the evening prior to the consecration vandals smashed many windows. The ceremony went on with temporary polythene windows but such was the local outcry, to a great extent sponsored by the workmen, that the villains were soon traced.

Vandalism on a house-building site usually falls off once some houses are occupied.

Some builders have drawn on this knowledge and taken it a stage further in that a suggestion of occupancy is created before any tenants arrive, by using temporary curtains and inside lighting.

Other builders who utilize their own workmen in caravans for watching purposes, move a caravan nearer to the almost completed building, or alternatively, have a temporary alarm, with a loud bell fitted. The question of a tethered dog being left on the premises may be considered, but obviously in this case all windows and doors or other means of access must be adequately secured.

Felt pens or aerosol paint sprays are a menace and one only has to consider the havoc caused on all motorway signs and public buildings to appreciate that building sites cannot escape attention. At the present time experiments are being carried out to find a means of erasing the various texts from brickwork, stonework, plaster, cement rendering, etc. and while some commercial acid cleaners get rid of the paint, these also damage the brick or stonework, etc. which is no answer on a new building. There is a case on record where, as a result of a paint spray effort about 1.5 metres (5 ft) from the ground on the gable end of a three-storey block of flats, the whole gable end had to be rebuilt, for the client refused to accept the erasure by acid or the possible insertion of new bricks at the affected part. Fortunately the offenders, three teenagers, were found and ordered to repay the full damages.

One must always report outbreaks of vandalism to the police, who now frequently operate anti-vandal patrols in their areas.

There is really no complete answer to combating vandalism as outlined, but additional watching staff, reward schemes, or flood-lighting likely affected areas, boarding-up or wire mesh are suggested. As with all security aspects, site management has to assess the cost of prevention against the cost of acceptance. One must admit that this is a poor state of affairs, but it is not restricted to the building industry and one can only hope that if the offenders are found, courts will severely punish in the hope that other potential vandals will be discouraged.

See Chapter 16 for the powers of courts to award compensation for vandalism.

11

Security of plant and tools

Plant

The hire of plant costs money and site management know this, but they should also be aware that in all cost-conscious and efficient companies a charge is made against the company's own plant issued to their sites. The reason for this is that a more accurate costing of the site is reached, and further it should encourage site management to reduce the time that a particular piece of plant is on their site.

In spite of these charges one factor stands out whether own or hired plant is used. There is always a degree of urgency when it comes to getting the plant on site to keep the job progressing, but there never appears to be the same degree of urgency to get the plant off hire and away from the site when no longer required, thus reducing charges.

The hire of plant is an expensive item and if site management is not constantly aware of the position a great loss can ensue. Because of these costs almost all companies require a weekly return of plant on site and how many hours it has been used. The site manager usually has to sign such a return, and even though it is completed by a site clerk, the manager should take the opportunity to reassess the position when signing the return.

Some site staff may feel that a charge for the company's own plant is not justified for it is a case of 'from one pocket to another', but they should realize that if they have an unwanted piece of plant on their site, when it could be required on another company site, the company is forced into extra expense of hiring for the other site.

Plant includes all machines, dumpers, hoists, compressors, generators, mixers, scaffolding, angle grinders, vibrating pokers and the like and every effort must be made to ensure that any plant whether own or hired is not damaged. Shovels, picks, wheelbarrows or similar articles are not usually the subject of a hire charge, but are regarded as disposable, but this does not mean that care must not be exercised.

It is absolutely necessary for all plant to be identified in some way, either by a serial number or at least daubed with company colour paint so as to discourage theft or misappropriation, but see later in the chapter for marking, etc.

Scaffolding

The National Association of Scaffolding Contractors have adopted a scheme of colour coding for all their scaffolding and scaffold battens and it includes fittings which can be very expensive. This scheme was necessitated by the actions of some scaffold-hire firms, who when collecting had a tendency to pick up everything from site, which was very often much more than they had supplied. The loss of scaffolding was suffered by any other hire company on site but, in all cases, the builder who had hired the scaffold, etc. was left to pay for the missing plant. Even with the colour scheme, the builder, unless he has a very keen checking on and off procedure, will be charged for any missing items and this causes endless aggravation with hiring companies. In his own interests, the builder must see that checks are made and signatures obtained.

It is appreciated that sometimes, on larger sites, it is necessary to use own scaffolding, and some hired, possibly from one or more suppliers, but this is not recommended for endless complications can ensue. This problem is eased somewhat if it is all suitably marked, which enables separate stacking for collection and counting. Often, when a collection is made there is a deficiency, which is a cost against the site, so just as there is an exact list, which should be checked when scaffolding is delivered, one must ensure that an equally exact record of that taken away is obtained against a signature. The practice of allowing companies to take away scaffolding, etc. not counted, must never be condoned, for the builder can be liable for charges which cannot be refuted.

It is appreciated that it is not unknown for scaffolding clips to be stolen, but there is no doubt at all that a large number are left scattered on site only to be buried in mud, etc. When not in use they should be stored in bins, not at the edge of the site.

It is pertinent at this stage to refer to the fact that most scaffolding work, whether erection or dismantling, is covered by a 'price' or 'bonus target' and it must be stressed that dismantling means the orderly, systematic lowering and neat stacking of all tubes, clips and battens ready for collection and transportation to the next job.

The rough practice of pushing over the scaffolding, thus bending tubes and leaving clips fixed to the tubes, is not satisfactory to qualify for full payments and site management must be firm in this respect. These observations do not apply to scaffolding subcontracted out, for all losses are the responsibility of the scaffolding contractor.

Dumpers

In common with all other mobile plant received on a site, it is accepted that, once the hirer has signed the delivery note, a presumption arises that the machine is in good working order, i.e. brakes and steering, etc. are efficient. In many cases there is no qualified man on site to do any checking but a licensed driver, i.e. a person who holds a driver's licence, may have sufficient practical experience to pass an opinion on the condition of the dumper.

It is recognized that by law, if the site boundaries are not exceeded, a dumper driver does not need a driver's licence, but he must be over eighteen years of age.

Site managers are well advised never to allow unlicensed drivers to use dumpers, for such persons are more likely to have an accident, to damage materials or even run into erected scaffolding because of the rear steerage factor alone. Managers must also be very careful in accepting verbal assurances of previous dumper-driving experiences with other firms.

Permitting all and sundry on a site to use a dumper should never be allowed, and it is good practice for all dumper drivers to be authorized in writing by site managers.

Passengers are forbidden by law to be carried on dumpers and no insurance cover is provided for any unlawful passengers. Any argument that passengers are sometimes a necessity to hold large bulky articles such as doors, floorboards, etc. on the dumper bucket, must never be tolerated. If the materials cannot be securely carried on a dumper without a passenger, then other means of transport must be used.

Mixers

Unless there is a central mixer of the 'pugmill' type it is advisable to move a mixer around and in all cases, whether it be the solid wheel or rubber wheel variety, to cover the wheels with an empty cement bag or cover them with sand, so that they do not become buried in mortar which sets. The removal of the mortar usually causes damage when the mixer is moved. At the end of the day mixers should be cleaned out (loose bricks and water are effective) and left with the bucket upturned, for this avoids the prevalent practice of having to hammer the bucket to remove mortar which adheres to it. This hammering can cause much damage, the cost of which will fall on the builder.

Holiday periods

Over the lengthy holiday periods as much plant as possible should be stored in the compound and, if this is not possible, the removal of all means of starting the engine, etc. must be ensured by complete removal of the batteries, keys, etc. Wheels can be made immovable by using chains locked to the chassis, the making of secret 'cut-out' switches and the use of draw-bar and lifting eye locks for compressors, etc. are advocated.

General

When not in use all plant must be immobilized, either by locking up or by taking away starting handles or ignition keys, etc. and, if it is possible the plant should be stored overnight in a compound. Mixers are no exception to this rule for if the starting handle is taken away, only a determined thief will be interested.

It is not unknown for vandals, frustrated by their inability to start machines left on a site, to put sand or like substance in the fuel tank with devastating consequences, and some efforts should be made to counteract this by having locking fuel tank caps.

Collection

At the time that plant is collected the site manager must get a receipt and check, so far as is possible, with the collector that the plant is not damaged, for so very often plant hire companies, when acknowledging return of their plant, add that a certain part was broken or damaged and such damage can be inspected at their depot within the next seven days. If a site manager has carried out the correct procedure, it could be advantageous to inspect the alleged damage, for mistakes are sometimes made. Problems very often arise when, after a request for a change of plant, or a cancel hire notification has been made, the hire company collects the plant out of working hours, sometimes when not even a watchman is present. Many such collections are followed by letters alleging damage or missing parts such as starting handles, compressor hoses, etc. for which a payment is demanded. Often the missing parts are available, having been stored securely and separately, in which case the hire company should be notified in no uncertain manner that the practice of out of hours collection will not be tolerated and the missing parts are available on site for collection *by them*.

Sub-contractors

Site managers must never hire or arrange to have hired plant for sub-contractors on site. Sometimes the sub-contractor's credit is low or he may already be indebted to the hire company. If the site manager takes on this responsibility, possibly on the understanding that the costs will be deducted from any amount due to the sub-contractor, it is ill advised, for generally the sub-contractor's payment is due well before the plant hire account is received. It must always be remembered that usually the sub-contractor has agreed to do the work for a fixed price and if the cost of plant has been allowed because it is required, it is his responsibility, and his alone, to arrange for it.

Sub-contractors should never be allowed to use company plant and site managers must never, out of compassion or other reason, permit it. All types of problems can arise, the worst being that the sub-contractor after constant usage may feel that the plant belongs to him. A case worthy of note occurred when a sub-contractor on drainage work had insufficient and ineffective shoring for a deep trench. The site agent agreed to the use of company timbers for this purpose, an accident occurred and the sub-contractor tried to wriggle out of liability on the grounds that the shoring was defective. These remarks are equally applicable to company shovels, picks, wheelbarrows, gumboots, wet clothing, etc. which should never even be loaned to a sub-contractor, who should supply his own.

The above restrictive use of company plant does not generally apply to labour-only sub-contractors, but it is essential that the site agent familiarizes himself with the minute details of the written contract otherwise in all innocence he can commit his employer to unnecessary expense.

If a sub-contractor has his own or hired plant on a site, it is his responsibility to provide adequate security for it.

There is a standard National Federation of Building Trades Employers (NFBTE) form of labour-only sub-contract agreement and this is readily available and should be prepared by the company and signed by the sub-contractor *before he starts the particular work.*

Tools

When a site is opened all companies provide numerous tools, the number and type being dependent on the size and type of work to be done, but all companies have the same experience in that those tools often disappear. The losses of tools, some valuable, are a major concern with all companies. While marking them, according to some opinions, will not stop them from being stolen by a deter-mined thief, it is a fact that if tools are marked not so many men will be tempted to steal. Sometimes, possibly out of weakness, site foremen allow lump hammers, shovels, wheelbarrows, axes, etc. to be used by sub-contractors' gangs of workmen, and many times because the tools are not marked they never come back – there is little that can be done for unmarked tools cannot be identified. The message is clear – all tools must be marked.

Consider the case of a thief arrested by the police and found to be in possession of a large quantity of building tools. Although it is strongly suspected that he may have stolen them, he denies it, stating that he has purchased them; he cannot produce any evidence of such purchase and as there is no power to force such evidence to be produced, the police cannot launch a prosecution on suspicion alone. It could well be that the police could contact a site where a theft of identical tools has been reported, but because the tools were not marked, the foreman can only say that the tools are simi-lar. This is not enough for court action, and ridiculous as it may seem, in such cases the 'thief' must be allowed to keep the articles.

The necessity for marking is well exemplified by consideration of thefts of pedal cycles in this country. Every year thousands are reported stolen and never recovered. Yet, over the same period, thousands of cycles are reported found. Common sense dictates that there must be a very close tie-up with the numbers, but the fact that only a few are identified is purely and simply because the losers did not know of the frame registration. There is no doubt that had these numbers been known possibly all the found cycles would have been identified. Various police forces have pointed out to the public the necessity and reasons for knowing cycle numbers and of the necessity to know of any identification marks on jewellery and household goods, etc.

The principle is the same and the moral is clear that all tools must have an identification mark.

Incidentally, it is no use having marked a tool unless a record is kept of the mark. This may appear very elementary but cases have occurred where the loser has told the police that a number is scratched on a certain part of the tool, but has been unable to supply the number. Initials are different.

Marking

Having accepted that it is common sense to have all tools marked to form a link with the proper owner, it is essential to explore the types of marks and the means of marking.

Names, initials, serial numbers or code numbers can be made on tools in several ways:

1. Stamping with a metal dye. NB: This method is not recommended for delicate measuring tools for the necessary force may interfere with the mechanism.
2. Engraving, etching, scratching as deeply as possible. NB: There are cheap engraving tools now available on the market, some operate by writing with mild acid and others by creating an electrical circuit. Obviously such etching, etc. must not be so deep as to damage any electrical insulation.
3. Burning wooden parts with a branding iron.
4. Painting with a known colour. 'Camrex' paint contains an identifiable pigment which can be traced to a known source.

Several larger and more expensive tools such as transformers, generators, even colt croppers, though marked, should warrant a signature when issued out to workmen on site, emphasizing that the person is responsible for the security of the article. The majority of men will accept this and ensure it is returned to the stores at night as opposed to leaving it on site, e.g. in a partially completed unlocked building. The system must be so tight that all are aware of it. There may be a school of thought that the job will be slowed down by any such system but this is quickly countered by the fact that the job may stop altogether if the particular tool is no longer available.

All thefts of expensive tools such as transformers and generators must be reported to the police who will always ask for the make and number or other identification marks so site management must be aware of these details. Through the efforts of CONSEC all thefts of mobile plant are placed in the National Police Computer for immediate reference if a thief is found in possession of any such articles. Many tools and pieces of plant have been recovered and thieves arrested through this system. A list of stolen plant and tools, and also a list of those in police possession for which an owner is requested, is regularly published by CONSEC. *If you don't know your number, it's no use.*

It is essential that all methods and means of marking plant should be registered with CONSEC for two special reasons:

1. To avoid duplication of marks, e.g. J.W.1, J.W.2, etc. could be either Joseph Wood Ltd or John Ward Ltd. If such a system of marking is contemplated it is advisable to check with CONSEC that it is not already in use, and
2. If marked tools are traced by the police, their first contact is with CONSEC, who if armed with details can short-circuit efforts to trace the owner.

Storage of workmen's tools

In accordance with National Working Rule 18.2, all employers, where practical and reasonable on a site, have an obligation to provide an adequate lock-up or lock-up boxes where employees can leave their tools, and furthermore, the employer is liable up to a maximum of £160 per claim for any loss by fire or theft of such tools which shall have been properly secured by the employee in such lock-up or lock-up boxes.

There are certain aspects of this rule which require more detailed analysis:

1. The tool store shall be so designated and the compensation clause does not apply to tools stored in any other place.
2. The rule only applies to directly employed operatives, and does not include labour only nor employees of sub-contractors.
3. This rule does not apply to plumbers and electricians

66

whose conditions of employment are not covered by these National Working Rules.

4. It could well be that for convenience, operatives, other than those directly employed, are allowed to use the tool store. It should be explained that the compensation clause does not apply to them.

5. The words 'properly secured by the operative' are specifically included and must be interpreted that in addition to leaving his tools in the lock-up an operative must also secure his own tool-box, etc. The reasons for the inclusion of this clause are clear in that in former days a workman's tools were inviolable, that is, no workman ever took the tools of his workmate, but regrettably such is not the case these days. Thus, if a man does not report for work, possibly due to sickness, and his personal tools are not secured, somebody may be tempted.

6. The maximum amount of £160 compensation is for each individual employee and so if the tool store is violated or destroyed by fire, and say ten employees lose their tools, the employer is liable to a maximum of £1600.00.

7. The employer must be satisfied that the tools claimed were in fact in the lock-up.

In all such cases, it is very pertinent to observe that if a workman reports that his tools have been stolen, he should be instructed to report the facts to the police, who will require full details in the form of a written statement. If a man should wrongfully and deliberately enlarge his claim by exaggerating the amount reported stolen, and such cases are not unknown, then any enlarged claim to his employer for compensation under the National Working Rules may amount to attempted theft by deception. Claims for compensation under this rule must always be very closely investigated.

There is a vital practical problem to which reference must be made and for which there is no easy answer.

If the tool store is violated and tools stolen, the discovery is generally made at the time of starting work. Immediately there is a 'hue and cry' from the workmen, who are all generally familiar with the compensation rule, for new tools to be supplied immediately by the employer so that the men can carry on working. Without tools, particularly if they are on bonus targets, they cannot earn.

When faced with such a situation, a site manager should always refer the matter to a higher authority for a decision.

Many firms authorize the purchase of the minimum vital tools for the job to continue immediately with a proviso, reduced to writing and signed by all concerned, that if the police recover the missing tools, then the newly purchased tools must be handed back to the employer.

If the stolen tools are never recovered the position is straightforward, but if they are recovered, then the new tools go to an embarrassed employer, who may not have any use for them.

Opportunity is taken to refer to the fact, borne out by statistics, that many cases of thefts of tools take place at lunch breaks where the workmen have left tools lying about where they were being used, and it should never be overlooked by site management that tools left lying about on site, when the site has closed, are often used by thieves forcibly to enter store sheds and site offices.

Workmen's clothing – fire

In accordance with National Working Rule 18.3 where an operative leaves clothing in accommodation provided by the employer as required by the Health and Welfare Regulations, e.g. drying rooms, the employer is liable up to a maximum of £30 for loss of such clothing by fire. Again this only applies to employees.

Before agreeing to the claim, which can be up to £30 per claiming employee, an employer must be satisfied that the clothing alleged to be burnt was placed in that place stipulated by him as the 'drying room', etc.

Surveying instruments

Theodolites, levels and the many other setting-out surveying instruments all have an identification mark, usually consisting of the type of instrument and the serial number. In all security-conscious companies these vital identification details are recorded, and further, these valuable instruments are only issued against a signature, usually the surveyor to whom it is entrusted or the site foreman/manager.

Some of these instruments are generally the highest valued single article on a site, possibly over £1000, therefore storage on site poses a serious security problem. Certainly they should be securely locked up when not in use and in many companies, unless there is a full-time watchman on site, the surveyor or site foreman is authorized, sometimes instructed, to take the instrument in his car for overnight security at his home.

12

Bonus targets

Payments

The price for a job, or a bonus target, covers nearly every operation on a building site, and it is commonplace for workmen to request the 'price' before starting the job. However, the one point which must be clarified is that the 'price' is for the satisfactory completion of the work, and *the site management decide if it is satisfactory*. There are many examples such as insufficient nailing of floorboards, insufficient screws in hinges, windows and doors not fitted correctly, faulty bricklaying, tiling and a host of others, but in all cases site management must not in any circumstances approve slipshod work, thus approving payment, for the possible repercussions are enormous. Sometimes other workmen have to be paid to rectify or even do the work again and, apart from the extra materials used at a cost to the employer, the trades following on may be delayed and the cost of subsequent maintenance greatly increased.

The most glaring example is revealed if one applies these principles to drain laying. If the work is done in a slipshod manner, it must be discovered by site inspection prior to back filling, otherwise re-excavation ensues, and the workman, whether sub-contractor or not, will not be pleased with having to do it all again at his own expense and will probably leave the site. The alternative, if done by other workmen, even if a countercharge is raised, is a great expense and loss to the company.

Thus keenness by site management is essential in the approval of work done, as it is also in agreeing to work which is only partly finished. Take the case of a tradesman or sub-contractor doing a specific job, e.g. plastering out a block of offices, factory premises or dwelling houses. At the appropriate time the work is measured by the surveyor or bonus surveyor who agrees that a certain percentage has been done, thereby agreeing to that percentage of payment. These records are signed by the site manager but before doing so he must be absolutely sure that the work has been done satisfactorily and that the amount is correct. It is not unknown, if overpayments have been made, for any subsequent lower correcting payments to be either disputed or result in the tradesman stopping work on the site, with consequent overpayments being required for other workmen to complete the work.

The exact location of the work done must be always identified on the written records, e.g. block number, unit number, etc. and this is no less important in records of cleaning out units, etc. prior to occupation. Sometimes local casual female labour is used for this latter work and then identification is vital.

Dayworks

Although almost all payments made are on the bonus target system, there are still cases where men, or a gang of men, are paid at an hourly rate, usually men of a ganger-labourer type.

It is not unknown for *extra* men to be shown on the ganger's return, and at times, either by accident or intention, this *fiddle* can be covered up by variations of the numbers of men on site. This system of having 'straw' men must be stamped out, and the only way is to insist that before starting work each day, each member of the gang 'signs on' or 'clocks in' if a time-clock is available. Even then periodic checks must be made that they are still on site, for if it is seen that site management is alive to the possibility of such a system of fictitious labour, temptation is reduced.

However, if any discrepancies come to light, particularly signing or clocking in for another, a criminal prosecution should be considered.

Another permutation of this type of fraud can arise on a big housing site where say a sub-contractor has been engaged for the joinery work and also given the joinery maintenance work on houses already completed. The foreman joiner must ensure absolutely that the maintenance work being paid for is not in reality the necessary remedial operations of work already done.

There are so many aspects of 'fiddles' of this nature which cost building companies large amounts that they are too numerous to mention, but can be restricted or prevented by the keen observation and firm action by site management.

13

Labour relations

If one includes in the overall security responsibility that site management conserves employer's assets, saves avoidable expense and prevents avoidable loss, then it is necessary for site managers to have some knowledge on the law and procedures applicable to labour relations, for ignorance can be a costly business.

Because they have a mandate to 'hire and fire' employees on their sites, they must be conversant with the National Working Rules applicable to the building industry. So very often in the past, apart from knowing that the Rules existed, it was not unusual to consult the appropriate union representative on site for advice and interpretation of these Rules. This is bad and in the long term can only lead to trouble.

Furthermore, failure to comply with the rather elementary procedure in disciplining or dismissing staff can attract very large compensation cash payments, which the employer can usually ill afford. Consequently a small blunder by site management, possibly due to ignorance, can have a devastating effect on any company.

It is often said that any organization must be administratively sound and nowhere is it more true than in industrial relations. Consequently it is imperative that every company circulates in writing to all site management the correct procedures for the employment and dismissal of labour.

Many companies issue Standing Instructions to site personnel covering the obvious aspects of site works, administration, etc. and opportunity should be taken to include the very important correct procedures covering discipline or dismissal in these or similar instructions.

The following outlines are suggested for inclusion.

Recruitment

The agent/foreman is responsible for:

1. Recruiting his own labour, possibly in conjunction with the supervisor (specific reference as to who places newspaper advertisements), and
2. Ensuring that all new employees complete, in writing, an application for employment form, which is sent to the office, where the written contract of employment will be completed, for issue to the employee.

Discipline

The agent/foreman is responsible for:

1. Day-to-day discipline of employees on site.
2. Issuing initial verbal warnings, e.g. bad timekeeping, standards of workmanship. If possible all verbal warnings must be witnessed, but in all cases detail of times and reasons must be recorded in the agent's site diary.
3. Issuing written warnings for persistent breaches of (2). NB: *The breaches of discipline must be generally of the same type and reasonably near in point of time.* For example, if a man is late for work in January, a similar breach in August with good timekeeping in the meanwhile, would not justify a written warning.

    Any written warning must be in simple terms. It must make reference to the reason for, and the date of, the previous verbal warning and details of the latest transgression which has warranted a written warning. It should be served in the presence of a witness.

Responsibilities (2) and (3) are prerequisites to any dismissal for minor breaches of discipline and must be observed, otherwise actions can be attracted for unfair dismissal.

Site management must appreciate that correct disciplinary control is of paramount importance and must be particularly careful when exercising this control to ensure:

1. Fairness can be proved.
2. Consistency.
3. Compliance with the law.

The interpretation of the legal provisions by tribunals provides the following guidelines on the subject:

1. The matter complained of must be clearly explained to the person concerned when the warning is given.
2. Warnings must be given in such a manner as to allow the matter complained of to be put right.
3. All final warnings must be in writing.
4. Before dismissal actually takes place, the person against whom the action is being taken must be given **the opportunity to state his case and explain the circumstances**.
5. Before dismissal, opportunity for appeals to higher management must be given and explained.

Dismissals

It is essential for all concerned to be familiar with the proper procedures for dismissal, for if action is successful for unfair dismissal tribunals take a very serious view if the rules are contravened and large amounts of compensation of up to 104 weeks' pay can be awarded.

However, employees with less than twelve months' service cannot take a claim to an industrial tribunal unless they consider they are being dismissed because of discrimination on the grounds of race or sex or because of some trade-union activity.

It is adequate for site foremen to have this basic knowledge on unfair dismissals, but because the law is quite complicated on this particular subject, site agents, etc. should always consult the industrial relations specialist in their company for detailed advice before any controversial action is taken.

Site agents must, however, be aware of the legal periods of notice which are required before an employee is dismissed.

Summary dismissal – periods of notice

1. *Up to first six days*. Two hours' notice to expire at the end of the normal working hours on any day.

   This is intended to cover a case where a man alleges that he is a tradesman/joiner/bricklayer, but before the six days has expired, the site manager realizes that the man is not a trained tradesman but is just a good DIY type and not suitable for employment.

2. *After six days but less than four weeks.* One day's notice to expire at the end of normal working hours on a Friday.
3. *After four weeks but less than two years.* One week's notice.
4. *Two years but less than twelve years.* One week's notice for each full year of continuous employment.
5. *Twelve years or more.* Twelve weeks' notice.

NB: Employees should give employers one week's notice except in (1) and (2) above, when the period will be as shown.

Instant dismissal for misconduct

There is no definition of misconduct but it certainly covers thieving, drunkenness, assaults on a foreman, etc. Witnesses are usually required but employees *must* be dismissed *at the time* that the misconduct comes to light. Later dismissal on the grounds of misconduct could well be the subject of legal action for unlawful dismissal.

Written reasons

Only employees with twenty-six weeks' service are entitled to a written reason for dismissal if requested, but in practice many companies, as a matter of policy, give a written reason for dismissal in all cases. Therefore, even though some managers may regard it as irksome, it is absolutely essential that they furnish the office from which the letters are sent with accurate details in writing of all dismissals on sites.

Redundancy

There is a whole series of legislation on this subject, but it is usual for this side to be dealt with by the administrators of the company. Suffice it to observe that if a tradesman is dismissed for lack of work he is being declared redundant, but of course does not qualify for a redundancy payment unless he has been employed by that company for 104 weeks. It could be said that there are more regular redundancies in the building trade than in any other industry, for men are usually taken on for a specific job, e.g. joinery work, and when that is finished they possibly move on to another company for similar work.

Employee leaving

If a man is sacked from a site, he is entitled to unemployment benefit, but if a person leaves of his own volition, in general terms he is not so entitled for six weeks. Site management must be constantly aware of this factor, for sometimes out of misguided compassion they come to an arrangement with an employee who wishes to leave to report that he was dismissed. This can have serious repercussions for the man may be later advised to institute action for unlawful dismissal and any change of story from an employer's written reply to the Department of Employment can be very embarrassing to say the least.

Absence from work

If an employee does not report for work and offers no supporting doctor's note or other reasonable explanation, he should be requested, in writing, to attend for work at a specific time with an explanation that if he does not, then it will be presumed that he has left of his own accord. The usual documents will then be sent to him, and as stated before, he will not qualify for unemployment benefit.

The foregoing is a brief résumé of procedures for recruiting, disciplining and dismissal of labour on site, but it is reiterated that the law and procedures governing labour relations are complicated and full of pitfalls, and therefore in all but straightforward cases, site managers should always refer the facts to their own company expert for advice and direction.

Fuller details of this complex subject are covered in another Construction Press publication – *Industrial Relations on Site*.

14

Sub-contractors

The methods of supervision and dealing with the materials of sub-contractors, whether nominated or otherwise, has been mentioned in various chapters, but it is such a vexed question, causing confusion and controversy, many times at cost to the main contractor, that specific reiteration is justified.

Materials

It is accepted that many of the following points are covered in the sub-contract documents but it is essential that:

1. The sub-contractor should have adequate arrangements for the receipt of his property. It is not advisable to allow company employees to sign for such property, for if it is substandard or deficient in number the company could have a moral liability. However, cases do occur when sub-contractors' materials, e.g. roofing tiles, are delivered to site before their workmen arrive, in which case there appears to be no alternative but reluctantly to give a qualified signature of receipt. In such an instance it is recommended that the sub-contractor be notified of the unsatisfactory system and informed that one of his employees must be on site to receive his materials.
2. A sub-contractor should never be allowed to accept and sign for company materials for the very obvious reason that he is rendering the company liable for payment.
3. The question as to who provides the store, who off-loads and who has access to the store should be fully explored and recorded before the job starts.
4. The location of the store for sub-contractors' materials must also be agreed, and if any security charges are necessary, the question of payment must be considered.

5. It must be made abundantly clear, in writing if need be, to the sub-contractor that he and his staff must comply with all security discipline on the site.

Plant

Similar definite arrangements must be made about plant in that only the sub-contractor must order his plant, never must a company order it on his behalf. The care of it is his responsibility, and under no circumstances should he be allowed to use company fuel for his vehicles or plant. It is risky to say the least to allow him access to the company's diesel store on the loose arrangement that he will pay for all the fuel used. This would mean supervision of supplies, records and endless 'hidden' costs to the company.

If, possibly to keep the project moving along, as a very last resort because the sub-contractor's plant has broken down, a site manager agrees to allow a sub-contractor to use company plant, there must be no loose verbal arrangements but a written agreement to accept the plant and a responsibility for any damage sustained.

Supervision

Site management are responsible for the project as a whole and therefore have a duty to supervise the work of sub-contractors and see that it is correctly carried out. The attitude that if work is not correct it will be the liability of the sub-contractor to rectify it is very shallow, for all must realize that it is not only the cost of rectifying the work but the hidden tangent costs which may involve the company.

Quite often sub-contract agreement specifies that the company has to provide certain facilities for the sub-contractor, e.g. the use of erected scaffolding for access and working platforms for roof tilers, but, if due to the laxity of the sub-contractor such facilities are extended beyond the time agreed, then charges must be levied against the sub-contractor. In the example given it could well be that the scaffolding may be wanted elsewhere, or if on hire, cancelled, and it is quite wrong for the company to have this expense thrust on it. Site managers have a responsibility to protect their employer's interest in these cases.

Payments

The site agent, when certifying the work done, must be sure that a sub-contractor's work is satisfactory in quality and quantity before approving any payments, *which should always be by cheque and not by cash.*

Because some companies, possibly very few, mandate their senior site agents to pay some sub-contractors, particularly labour only, direct, and this system is not recommended at all for obvious reasons, reference must be made to the Construction Industry Tax Deduction Scheme.

Briefly, if the sub-contractor holds and produces a sub-contractor's tax certificate (Form 714) issued by the Inland Revenue, the contractor pays the sub-contractor in full. If, on the other hand, the sub-contractor does not hold, or if he does, does not produce, such a certificate, the contractor must make a tax deduction (currently 35%) from the payment and pass such deduction direct to the Inland Revenue.

There are three types of sub-contractor's tax certificate: Form 714C for companies, Form 714P for partnerships and Form 714I for individuals. The holders of exemption certificates 714P and 714I must give the contractor a completed special voucher (Form 715) for every payment received without deduction.

The above are the basic requirements to be observed, but fuller details of the operation of the scheme are covered by booklet I.R. 14/15 (1977) obtainable free of charge from any Inland Revenue office.

Labour only

Before the job starts all labour-only sub-contractors must be served with and sign a yellow form agreement, Form No. C/20, obtained from the NFBTE. This gives exact details of plant, etc. to be supplied and also details his responsibility for security, safety, insurance and many other vital aspects.

A case is recalled when a drainage sub-contractor had agreed to supply trench sheets and shoring timbers, but his liability had not been reduced to writing on the above form. Because the timbers, etc. were not forthcoming, the site manager, in furtherance of expediting the work, supplied them from company sources. A serious accident occurred when these timbers collapsed due to faulty fitting, and this resulted in intense debate and argument as to which insurance company was liable. It is always advisable to have all arrangements and agreements with sub-contractors reduced to writing, for no matter how friendly and co-operative procedures are at site level, possibly due to working amicably together for long periods, such relationships tend to become strained if a large insurance claim is in dispute between different insurance companies.

15

Liaison with police, fire service, public, schools and clubs, and prosecutions

Police

Reference has already been made to the amount of knowledge, assistance and advice that builders can get from the local police, particularly the Crime Prevention Department, and the means of obtaining contact will now be explored.

Calling at the local police station can be very frustrating, for sometimes difficulty can be experienced in even getting further than the counter clerk, who may or may not be able to fulfil all requirements.

Initially, a letter on the lines shown at the end of this chapter should be sent to the officer in charge of the local town police station. It will be redirected to the station which provides police cover for the area of the site. NB: *In companies which employ a security officer, it is usual for him to send the letter, but on his visit to the police station, he should be accompanied by the site agent.*

This letter should be sent before the contract starts and lots of information can be gleaned from the subsequent visit, e.g. local crime picture, extent of local vandalism, experiences of previous builders in the district, what they did about watching, reliability of local security firms and other facets not connected with security such as local labour, tipping facilities, etc.

When the job starts the Crime Prevention Officer, as a result of the initial contact, will no doubt visit the site and make recommendations on security aspects. He will be experienced in the difficulties of securing temporary accommodation, such as site offices and stores, and will not recommend very expensive systems which would make each hut a fortress.

It cannot be stressed too much that police co-operation is a great advantage, for they can arrange help in many ways and the ideal approach is by the initial letter.

A case readily comes to mind where a tender was being estimated for the building of a factory. Police contact revealed that the site had previously been occupied by a bakery, which because of the number of burglaries and the fact that the insurance premiums were so large, had been forced to close. Obviously a rough area. Consequently, a large sum was included in the tender for a full, strong perimeter fence. The tender was successful and in the space of nine months to the completion of the work, not one break-in was experienced. The moral is clear, for in this particular case even the police were amazed.

Fire Service

Very often the location of a site has an address which is not well known in the locality, particularly on new industrial or housing sites, where sometimes in a green-field situation the location is only indicated by a numbered field on an enlarged Ordnance Survey map. In this, and all other cases, it is always advisable for builders to notify the local fire service, in writing, of their presence in the area, immediately the site office complex is in being. A sample letter is given at the end of this chapter.

The fire service will be very appreciative of such a communication and a representative will visit the site to assess the nearness of the water supply – a vital factor, for site-office fires are notoriously fast burning – and record the ideal line of approach for their vehicles, particularly before the roads are formed and laid.

Many advantages can follow a close liaison with the fire service, whose officers will readily give advice on site fire precautions, location of petrol/diesel tanks and many similar matters.

Schools

Liaison with local schools must always be made, particularly if the site is adjacent to the school, if the site has been used as a playground by local children or if children have regularly used the site as a short cut *en route* to and from school.

Initially it is advisable to send a letter on the lines shown at the end of this chapter.

Experience has shown that advantages and co-operation can be obtained by the site agent (or the security officer) visiting the schools and giving a short talk as to how buildings are made, with reference to the functions of the various trades and the order in which they operate. Children usually find this most interesting and a change from routine education. A short supervised site visit by children is also very helpful.

Many cases can be recalled where children have given information on thefts/damage on sites where they have had talks and visits.

In one case a pair of bungalows ready for handing over, merely awaiting acceptance by the clerk of works, was forcibly entered, a fire was started in the roof space and efforts to douse it so damaged the water system that both dwellings were flooded. The culprits were not traced until the headmaster of the local school, after a promise of confidentiality, gave the names of the three teenage offenders which had been passed on to him by two junior scholars.

Sometimes the psychological approach is fruitful; for example, after a spate of trespassing and damage on the site, the site agent can have a word with the 'leader' and seek his help in keeping children off the site. This approach often works.

It is a very commendable policy, having been used many times, for any company, having sought the co-operation of schools on the above lines which resulted in a virtually vandal-free site, to consider the presentation of a small gift towards a school project. The school will appreciate it and information about such generosity will soon be widespread.

Youth clubs

Similar letters sent to local youth club organizers may have the desired effect but experience reveals that success is not as common as letters to schools. However, it is at least worth an attempt at co-operation.

General public

The cooperation of the general public is invaluable for the security of a building site, and the author cannot begin to estimate what thefts and damage have been prevented over at least fifteen years by such co-operation. While it is accepted that there is an attitude in some people not to get involved, the majority of members of the public are aghast at the spate of thefts and vandalism which occur generally and are only too willing to co-operate, more particularly if anonymity is promised.

It is because of this attitude that it is imperative at the time the site work is started to deliver a circular letter to all dwellings which are adjacent or overlook the site. Some will fall on stony ground but others will have the desired effect. See the sample letter at the end of this chapter.

It may well be that some builders will fight shy of a reward of £50 but the vital words 'up to' should always by used for information of small cases of theft. It is not unknown for employees to be caught as a result of this reward scheme.

The letters must be supplemented by large posters indicating that a reward scheme is operating fixed on compound gates, site offices, etc. They should be pasted to hardboard or similar material and covered with clear polythene for protection against the weather, in which case they can be recovered for use on other later sites. The posters must be kept simple and to the point.

New tenants

Very often on large housing sites when some dwellings are occupied while work is still in progress, new tenants are tempted to take materials found on site for additional paved paths, extra shelving, etc. to their new dwelling, and a letter should be given to them at the time of their occupation of the new dwelling warning them that such activities are unlawful.

This system has been used with success in that so far as can be ascertained many who would have been tempted have been 'warned off' by the letter.

A sample letter is shown at the end of this chapter.

Empty premises

Reference must be made to the following case where letters to surrounding houses were a complete success. Due to rationalization, a joinery works on an industrial estate was to be closed. There were many empty premises on the estate which had been broken into and vandalized almost out of recognition: doors and windows had been smashed and many wire compound fences pulled up and made useless.

A letter, reproduced at the end of this chapter, was distributed, and during the eighteen months it took to sell the premises not one pane of glass was broken and no damage or break-in occurred, even though no regular watchman was employed. This must be a tribute to seeking and getting public co-operation.

Thefts or damage

If a theft or damage of any magnitude occurs, the facts must be reported to the police and details forwarded in writing to the office. It is appreciated that the word magnitude is rather ambiguous, but clearly the police need not be told of say one broken window, not part of a break-in, nor if the foreman's pencil has disappeared.

However, for the police to get a proper picture of what is happening in their area, major thefts and damage must be reported to them, for it is not unknown for thieves or vandals to be traced and admit other offences of which the police have no knowledge. This often causes extra work in identifying property traced to the offenders, particularly if the builder has completed the project and left the area.

In all cases of reports to the police it is necessary to confirm the complaint in writing in the form of a statement, which the officer will take down.

Restitution and compensation

Generally speaking, the police are called in because all complainants either want their property back or adequate compensation, and in this connection the following observations are made:

1. If a person pleads guilty or is prosecuted to conviction, in addition to any fine, he can also be ordered to recompense the complainant up to £1000 for each charge whether it be theft or damage.

2. This recompense can be either the return of the stolen goods or their monetary value. If the goods are recovered intact, for example, a brush, shovel or other like article, the procedure is straightforward, the complainant will get the property back; but if stolen bricks have been used to build a wall or stolen kitchen units have already been fitted, then obviously they are useless as such, if recovered, to the builder.

   In any statement given to the police it should be clearly explained in those latter cases that it is useless for the police to dismantle the wall or unfix the units and that monetary compensation is preferred. The police are very familiar with this procedure and magistrates are generally very understanding about such 'damaged' stolen goods. However, while the law states that on a conviction the property

85

is automatically returned to the loser, if monetary compensation is desired on the above grounds an application to the court must be made by the loser. To avoid unnecessary court attendance, particularly in guilty pleas, it is essential for the statement given to the police to include the words 'in the event of a thief (or vandal) being found guilty I hereby authorize the police or their prosecutor to make application on my behalf for compensation'.

3. The value of the article stolen and given to the police must be the cost of replacement, i.e. a true value, and possibly not what was paid for it after various discounts for bulk buying, early payment, etc.

4. In cases of damage the amount must be that of repairing and replacing the article damaged; for example, the cost of a broken window must include the cost of the glass plus the cost of a workman to take out broken pieces and refit new glass. Many times, the cost of the labour greatly exceeds that of the material. Damage also includes consequential damage, e.g. interference with the water system which causes the flooding of a parquet floor, the pieces of which are dislodged and distorted.

NB: *Although there is a legal provision for the award of up to £1000 compensation against persons convicted of these offences, so far as is known there has never been action taken against a thief or vandal for this to include the costs of the delay to the project, but it may be worthy of legal consideration for one can imagine cases where the theft or damage could seriously delay a project with consequent expense, e.g. theft of or damage to a specially made font for church premises.*

It is quite lawful for site agents to accept payment for stolen goods from the thief, already *en route* to being prosecuted, or damages from a vandal in a similar position, but the latest development must be notified to the police. In no circumstances must any promise be made, after the facts have been reported to the police, that there will not be court action. Once cases have been reported to the police it is their decision as to whether or not a prosecution ensues.

Reference must be made to the question of alleged losses over a period, for example at the beginning of a project to build say 20 houses, each having 20 identical inside doors. Four hundred will be ordered by the buyer, but towards the end of the job it is realized that a further 10 doors are required to complete. Because they were signed for it is presumed that they were counted on receipt and because there is no evidence of damaged ones on site, neither any evidence of a breaking into the door store, some managers presume that over the period 10 doors have been stolen, but the police will not take kindly to such complaints of theft. Rightly, they observe that there has been some slackness by the management either by bad checking, bad stock control or allowing thefts to go unchecked.

Court attendance

In many cases where guilty pleas are entered at court, the attendance of the complainant, usually the site foreman, is not necessary, but if a plea of not guilty is known or anticipated, the police will warn, either verbally or by letter, the person who gave the statement of complaint to attend court at a specific time and date. This procedure is commonplace and the service of a witness summons is not required, unless the complainant openly refuses to attend without one. It is very rare for witness summonses to be required.

It is customary for witnesses at court to be soberly and properly dressed and except in very rare circumstances they will be required before giving evidence to swear before God to speak the truth. The prosecuting solicitor will conduct the witness through the 'examination in chief', followed by the 'cross-examination' by the accused or his solicitor, followed by the 're-examination' by the prosecutor on any matters raised in the cross-examination.

The judge, jury or magistrate assess the value of any witness's evidence on his knowledge of the facts, his obvious truthfulness and lack of bias in his testimony.

Many persons in the construction industry are reluctant or do not relish attending a court and giving evidence, but usually there is no cause for anxiety or worry if the true facts are related. Courts are not unmindful of the effect of giving evidence on the nerves of witnesses and make due allowances for it.

If a site manager has to go to court to prove the charge, he is entitled to any wage losses, if such be the case, and to any travelling or subsistence expenses because of his court attendance.

It is at the time of possible cross-examination of his evidence that his efforts at site security are rewarded for if he has been slack in overlooking previous misdemeanours or allowed debatable 'waste' to be taken from site, he can have a rough time and the charge could be jeopardized. There have been many cases where charges of theft have been dismissed on these grounds.

Very often difficulties are experienced when the police recover and hold vital materials, special plant, etc. as evidence to produce in a court, and until it is released the job either stops or if a replacement is obtained so that the job can continue, the materials and or special plant will be useless for future work. In this case, and in cases where large quantities are stolen, the police are usually very co-operative in having the exhibit photographed, thus releasing the material for instant use, and producing a photograph in court.

Sample letter to Fire Service

Name and address of company
Tel. No.
Date

Dear Sir,
(Address of site)
We have been awarded a contract to build (or we are building) houses/factory at the above address. We have now completed the erection of our office complex, and while we are not hoping for an emergency, we feel that you would like to know of our existence in your area and our exact location.

You may be interested to know that one of our priorities will be to arrange for the site roads to be available as soon as possible to allow movement of vehicles around the site.

We would appreciate the advice of your officers on our fire precautions if arrangements could be made for a site visit.

Yours faithfully,
for and on behalf of
(name of company)

Site agent/Security officer/Adviser

To: The Officer in Charge
Central Fire Station
(name of town)

Sample letter to Police

Name and address of company
Tel. No.
Date

Dear Sir,
Building site at . . . . .
We have been awarded a contract to build (we are proposing to start building) houses/factory at . . . . . . in your area. It is hoped to start our operations on (date), and our resident site agent/foreman will be Mr . . . . . ., who we know you will find very co-operative. We expect to complete the project by (date). We are a very security- and safety-conscious company, and we shall make every endeavour to safeguard our materials, etc. against theft or damage, and in the near future the writer or his representative will visit your station and will be very pleased to discuss with your officers any additional measures which we can take, having regard to local conditions.

We would like to assure you of our policy to assist and support the police in any criminal proceedings which you may contemplate, and would like to assure you that it is not our intention to report very small cases of theft or damage, unless there is a reasonable chance of detection.

Attached hereto, for your information, is a copy of a circular letter which we shall deliver to every dwelling house which overlooks the site.

Yours faithfully,
for and on behalf of
(name of company)

Site agent/Security officer/Adviser

To: The Officer in Charge
Central Police Station
(name of town)

Sample letter to Schools

Name and address of company
Tel. No.
Date

Dear Sir,
You are no doubt aware that this company has been awarded a

contract to build...... in......, a site adjacent to your school, and that we have recently commenced operations.

Every effort will be made to inconvenience the local people as little as possible and it is with this in mind that we write to you personally to apologize in advance for any noise, etc. which may possibly interfere with your staff during their teaching duties. We feel sure that you will find our Mr ......, who will be in charge of the site, very co-operative.

We are a very safety-conscious company and you will readily appreciate that in the course of any construction work there are times when temporary erections and excavations can be dangerous. This is particularly applicable to trespassers on the site who may not be aware of the hazards. It is our sincere desire to avoid unfortunate accidents and damage to our work that prompts us to write to seek your co-operation in stressing to your children that this or any other construction site is not a playground. We certainly understand that during evenings, weekends and holidays the children are not under your control, but it could well be that a general word from you to the children would avoid any unfortunate accidents. We do attempt to enlist the help of parents on similar lines as the attached circular letter which is delivered to all houses overlooking the site, shows.

We hope that you will co-operate with us in this small, but vital, effort to avoid any accidents and damage on our site.

> Yours faithfully,
> for and on behalf of
> (name of company)

The Headmaster

Site agent/Security officer/Adviser

Sample letter – New site – surrounding tenants

> Name and address of company
> Tel. No.
> Date

Dear Sir/Madam,
This company regrets any inconvenience to you and your family by their building operations near to your house.

The building of and repairs to houses, etc. necessitate, of course, making excavations and erecting temporary structures which are made as secure as possible, and you will readily appreciate the possible dangers if children are allowed to use the sites as a playground. In addition to the danger, children can do excessive damage, with consequent financial loss to the company. Please, therefore, for their safety, do not allow your children to trespass on the site, as you are legally responsible for their conduct and activities.

We must further draw your attention to the fact that no person has the right to take anything from site, even though it seems to be of no value, and it is the policy of this company to report all such unauthorized takings to the police with a view to criminal prosecutions. In this connection, the company agrees to pay up to

FIFTY POUNDS CASH

to anyone giving information, *which will be treated in absolute confidence*, which leads to the conviction of any person for theft or damage. If you see any vehicle possibly involved in the removal of materials, please take the registration number and then let us know.

Yours faithfully,
for and on behalf of
(name of company)

Site agent/Security officer/Adviser

Sample letter – New Tenants on estate

Name and address of company
Tel. No.
Date

Dear Sir/Madam,

You are a new tenant of a house on this estate, built and completed bv this company to the architect's specification and requirements. It may be to suit your own convenience that you feel that extra shelves, garden paths, etc. may be advantageous, but we must stress to you that you have no right or authority to take any property from the site. We have in mind pieces of timber, odd bricks or similar materials.

It is the policy of this company to report all such unauthorized takings to the police with a view to a criminal prosecution, so in your own interest please do not take anything from the site.

Yours faithfully,
for and on behalf of
(name of company)

Site agent/Security officer/Adviser

Sample letter – Empty premises

Name and address of company
Tel. No.
Date

Dear Sir/Madam,

You will be aware that we have ceased joinery production at our adjacent factory site, and to prevent thefts from and damage to our premises certain security measures have been taken.

These precautions necessitate the continued use of live electric cables, etc. and in the interests of safety we hope that you will discourage all children from trespassing and playing on our property.

We should be most distressed if anybody sustained any injury while on our property and this letter is prompted by a desire to prevent any unfortunate accidents.

It is the strict policy of this company to report all unauthorized takings from, trespassing on or damage to our property to the police with a view to criminal proceedings.

In this connection, the company agrees to pay up to

FIFTY POUNDS CASH

to anyone giving information, *which will be treated in absolute confidence*, which leads to the conviction of any person for theft or damage. If you see any vehicle involved in the removal of materials, please take the registration number and let us, or the police, know.

Yours faithfully,
for and on behalf of
(name of company)

Site agent/Security officer/Adviser

16

Conclusion

Efforts have been made to cover the many factors of security on building sites in its widest sense. They include new buildings, blocks of flats, houses, factories, offices, etc. and the modernization of existing buildings. They cover projects from the pre-planning stage to buying, receiving, documenting, secure storage, issuing to workmen and the use of materials to the many aspects of reducing the possible liabilities which may result from slack or indifferent site management.

However, there are a few other matters to which reference must be made.

It is an accepted fact that untidiness and carelessness invite petty thefts and damage, and managers should appreciate that it is simple economic sense to safeguard their site and stocks so as to emerge with a minimum wastage by theft or damage. Losses by wastage alone can be astronomical, and many site managers agree that on an average-sized site the employment of a conscientious labourer merely to roam about the site picking up abandoned and half-buried items is economically justified. One only has to think of scaffold clips, tie wires and tie irons, damp course, half-buried bricks, rubber boots, industrial gloves, wet suits, shovels, picks, road pins, and the like. The list is far from complete as all experienced staff will know.

Some management erroneously rely solely on insurance to cover losses, etc., but everyone must be aware that nearly all premiums for insurance against thefts on a site are negotiated annually or specifically before a project is started. Usually the builder is responsible for the first stated amount of any loss, but the main basis for negotiating insurance premiums is the previous claims record.

Site managers must liaise very closely with the client either directly or through his architect or clerk of works and accept his advice on the building aspect, and to insist, if such advice or direction is outside the scope of the original contract, that variation orders or on-site instructions are requested in writing so that the necessary payments can be obtained for the additional work involved.

Another aspect of security which must always be rigidly adhered to can arise if the building project is an extension to an existing building or a new building in a particularly security-conscious areas. In these cases any security restrictions or systems in operation in the existing building will be communicated by the client to the construction company and the site manager, who together have a clear responsibility to see that all the men on the site are familiar with the existing security restrictions.

Two instances come readily to mind:

1. A new wing at a mental hospital prompted a letter from the client giving details of legal restrictions. It was considered advisable to give every employee, including sub-contractor's men, a copy of the notice which is reproduced in Appendix 2.

2. A new store to be build within the confines of an existing factory where valuables were made and where the perimeter was sealed and, with union approval, a search clause for all employees was operating. The builder was notified and having accepted that the searching would also have to apply to his own employees, notified all his men in writing of the likelihood of a search adding that only men agreeing to this procedure would be sent to the site. The men appreciated that the search was necessary because of the high value of the products and the ease with which they could be transported. Because the men had correctly been notified and given the reasons for the searching prior to going to the site, it was accepted without query. Had the site manager not been so efficient, all sorts of problems, e.g. strikes, actions for breach of, or delay in the completion of, the contract could have ensued.

Site management must accept unreservedly that site security in all its aspects is their responsibility and even though security advisers are engaged it does not absolve them from that duty. To assist them in this matter, a Security Code of Practice, used for many years, is reproduced in Appendix 3.

It is no good site managers being very security conscious if they do not attempt to educate their staff into a like security frame of mind.

It must again be stressed that security is not only the provision of locks, bolts and bars, but the preservation of the employer's assets and the prevention of losses to the employer's detriment.

Many references have been made to wastage on building sites, and it must be accepted that much of this loss can be avoided by careful neat storage and the application of common sense.

The theory that the maximum impact on any subject can be achieved by involving as many senses as possible prompted the NFBTE to produce a series of four anti-waste posters which are used on sites by many building contractors.

One cannot estimate the good results which will become apparent if site managers educate and thus get support of their workmen in seeing that every site is an example of tidiness and neatness and thus good for security.

Appendix 1

Anti-vandalism practices in the building industry

(The author gratefully acknowledges permission to reproduce this extract from Appendix F of the publication Protection against Vandalism *by the Home Office Standing Committee on Crime Prevention.)*

Vandalism on building estates in the initial stages is created partly by design features which attract, excite and offer opportunities for uninhibited play, adventure and vandalism. Lighting fittings of unusual design, low roofs and others with adjacent walls offering easy access, blocks of garages, sheds, etc., hidden from view, signposts and fittings which invite climbing and swinging, projecting garage door handles at knee height, recessed wall lights at kicking height, are examples of such features and, once commenced, initial damage is usually continued and becomes increasingly difficult to prevent.

In later stages lack of thought in design and layout often results in damage to fences, low walls, ornamental grassland, flower beds, etc., by tenants taking short cuts and making routes (which are inevitably followed by others) where none was intended.

Anti-vandalism precautions should be considered in advance by architects, builders and all who are concerned with building construction. These should include considerations of layout and design, choice of materials and finishes less susceptible to damage, and the use of technology.

The following is a selection of anti-vandalism practices employed by some sections of the building industry, many of which provide the added advantage of security against theft.

Preventive measures

Perimeter fencing installed of a type which permits exterior inspection of the site with attention to strength, height, fitting (to prevent crawling under) including appropriate types of gates and restriction of unnecessary entrances/exits.

Separation of site offices, sheds and materials, etc., from fences to reduce opportunities for damage from exterior and climbing. Securing of all stores, offices, sheds, etc., during non-business hours.

Avoidance of unnecessary and excessive storage of materials on the site with as little as possible stored in the open.

Newly constructed garages fitted with locks and used for storage purposes.

Security lighting installed to create deterrent effect.

Immobilisation of all mechanical plant to prevent operation by intruders and storage in locked compound where possible.

Insistence on orderly, uniform stacking of materials where practicable to create appearance of care and attention and to make initial acts of damage immediately obvious.

Installation of intruder alarms, particularly in premises in which fragile or expensive property is stored, or with a history of vandalism.

Illumination of diesel tanks and re-siting in positions readily visible to the public and with adequate locking devices.

Selected personnel, or private security organisations, made responsible for checking the security of all buildings, gates, compounds, etc., within the site on termination of business hours.

Glazing of windows delayed until near-completion of the building.

Introduction of security training programme to combat theft and vandalism with written security instructions binding upon management, staff and sub-contractors.

Publicity by press, placards and letters to householders emphasizing the problems of vandalism and offering rewards for confidential information on those responsible.

Glass substitutes

Glass is probably the most fragile building material in use and is, of course, particularly vulnerable to acts of vandalism. Many companies are engaged in the production of glass substitutes with high resistance to damage. The versatility of these products enables widespread use to be made of them by the building industry. The main virtue of these organic glasses is perhaps in their lightness and strength, although there is some susceptibility to scratching and dust collection and the effect of ultra-violet light. The following are examples of glass substitutes currently being used in the building industry:

Acrylic, Glassfibre reinforced Polyester, Polycarbonate, PVC.

Protection of surfaces

There has long been the need for a surface-coatings system which would be durable, resistant to damage, and impervious to crayons, pencils, felt pens, etc. One such system incorporates hard aggregates, bound by special resins and overcoated with abrasive-resistant finishes. This product is sometimes used in the building industry as protection against vandalism in public lavatories, subways, bus shelters, etc. Conversely, there is the need to avoid soft-textured wall finishes which invite scratching and defacing, particularly if of a colour contrasting with the substrata.

Anti-vandal materials and fittings

There appears to be an increasing interest by the building industry in the use of various types of materials and fittings with anti-vandalism in mind. Typical of these are heavy-duty stainless steel sanitary ware in bowl and stall urinals, W.C. pedestals, wash basins, etc., vandal-resistant glazed ceramic partition blocks for use in public conveniences, public baths, etc., vandal-resistant battery garages, etc. There is also a continuing development of anti-vandal light fittings available to the building industry involving the use of glass substitutes, concealed fixing screws and inaccessible interior fittings.

It is difficult to predict the degree of wilful damage which will develop during and after building construction. Preventive measures may not be adopted unless a high incidence of vandalism is anticipated. If, however, the risk and the remedies are kept in mind from the design stage onwards, it may be possible to build in protection at little extra cost or loss of amenities.

Appendix 2

Regional health authority
Notice to building contractors

Workmen shall avoid, *as far as possible*, any contact with any patients in the hospital and particular warning is given of the following offences which are punishable at law, viz:

(a) Supply of intoxicants or drugs to or for any patient.
(b) Ill-treatment of patients (Section 126 of Mental Health Act, 1959).
(c) Sexual intercourse with a person receiving treatment for mental disorder (Section 128(1) (a) (b) (3) of Mental Health Act, 1959).
(d) Assisting patients to absent themselves without leave, etc. (Section 129 of Mental Health Act, 1959).

Workmen shall not purchase from, nor sell to, any patient any article whatsoever.

Workmen shall not accept from any patient any letter or other communication whatsoever for delivery either by hand or by post.

No patient shall be allowed to ride on any form of mechanical vehicle.

The foreman or other person in charge of workmen shall ensure that all tools, ladders and equipment of any kind shall be stored when not in use in such a way as to be free from interference by patients.

The foreman or other person in charge of other workmen shall ensure that all vehicles are properly immobilized when not in use, and made secure from interference by patients.

It is the responsibility of the contractor to see that all his workmen are fully aware of, and understand, the above rules.

Appendix 3

Security Code of Practice for Site Managers

A. Introduction

1. Losses caused by crime, carelessness and wastage in the construction industry cannot easily be assessed but *will show* in the final balance.

2. Untidiness and carelessness invite petty thefts and damage.

3. It is simple economic sense to safeguard your site and stocks and emerge with a minimum *wastage* by theft or damage.

4. Insurance companies are raising their premiums yearly and *always remember* premiums are based on claims experience.

5. *Damage, theft and wastage* can be prevented by
   (a) precautionary action;
   (b) careful thought;
   (c) physical means (static security).

On economic grounds alone it is impossible to make all stores a fortress and in any case determined thieves are difficult to counteract.

Basic security precautions are vital.

B. The cost

1. Savings cannot be assessed in cash for it means estimating an unknown possibility, but any company which has taken security and safety seriously and given it thought, particularly in educating their employees on the subject, *has never regretted nor abandoned the policy.*

2. It is dangerous for any company, or its employees, to convey to clients an air of indifference to vandalism and theft.

3. *Prevention is better than cure*, and the only criterion for comparison is an assessment of the situation should the entire police service of this country be removed for six months.

This theory must be applied to your site, bearing in mind:
- (a) difficulty, delay and cost of replacing materials;
- (b) general disruption that follows damage and/or theft;
- (c) loss of profit and client's goodwill;
- (d) financial loss – uncompleted or delayed contracts.

You must
- (a) remember *prevention*;
- (b) liaise with local police;
- (c) practise and encourage others in *war against waste*, however caused.

C. Depot

1. Stocks must be kept to a minimum – buyers to be kept conversant with progress and requirements and depot stocks.

2. All materials received to be physically checked.

3. Materials to be safely stored and proper records to be kept. *Remember* bad storage encourages waste and damage.

4. The use of burglar alarm systems should be explored but they are expensive, and consideration must be given to value of stores and anticipated losses should there be no protection.

5. Proper records to be kept of *all* outgoing stores – only that required within a reasonable time to be sent to sites. Remember sites are generally more vulnerable than depots, but common sense should prevail when assessing requirements. NB: The job should not be held up for materials.

D. Site security

Each site must be considered as a separate security problem. Start from perimeter and work inwards to a central compound where most valuable property is contained.

1. Perimeter fences – if considered necessary

Various types can be used – close boarded, paling, wire, sheet metal – governed by local conditions, but remember *a perimeter fence is only as strong as its weakest point*. (If broken secure adequate replacements.) NB: Materials should *not* be stacked against perimeter fencing, it enables thieves to climb over, thus defeating the object of the fence.

2. Entrances

(a) Number *must* be restricted, the fewer the better.

(b) Gates must be substantial and *not* capable of being lifted off.

(c) No gaps under gates or fencing under which people could crawl.

3. Site offices

(a) Must not form part of the perimeter fence – ideally there should be a 1 metre (3 ft) wide passage between.

(b) Must be as near to site entrance as possible for checking and directing deliveries.

(c) If possible checker's cabin to be elevated to check that vehicles are empty when leaving.

(d) *All* cabins should have proper locking-up facilities and window shutters which must be secured when site is not working.

4. Store sheds and materials compound

(a) As much as possible to be stored inside but value and cost of transport must be considered.

(b) Consider use of garages, etc. when built, for storage when they can be made secure.

(c) Stacking must be *neat and tidy* – takes up less room and theft or damage is immediately obvious.

(d) Timber must be stacked on scaffolding racks and covered.

(e) Storage of workmen's tools must be in a locked building (National Working Rule 18.2).

(f) Sheds must be sound and adequate for the safe storage of materials and erected on levelled ground.

(g) Effective measures must be taken to repair damaged sheds either on site or at the depot.

(h) Metal sheds which are used for very valuable property must be closely examined for defects, particularly the roof sections which must be bolted down.

(i) Screw heads which show on the outside should be 'burred'.

(j) The use of bolts, with nuts on inside, is advisable for all hasp and staple fastenings. NB: An expensive lock is useless if good screw heads are visible.

(k) There must be adequate discussion with sub-contractors on the storage problem. In general they will agree to the general pattern and consideration must be given for adequate protection of their valuable stores.

5. Lighting

Use of adequate lighting must be considered, particularly floodlights which are a known deterrent to vandals and thieves – a time-clock is a necessity.

6. Parking

No employee must be allowed to park anywhere on the site – there must be provision for *all* cars to be parked together – a natural deterrent.

7. Cash

Must *never* be left on site.

8. Action after theft or damage

(a) Inform local police *immediately*.
(b) Memo to head office for security department.
NB: It is essential that all tools and portable property should be marked so that positive identification is possible; if not, even if obviously stolen property is found on a suspect, it may have to be handed back to him because of lack of any evidence of ownership.

9. Watchman

Professional security firms are expensive for full-time cover and, if possible, site labour should be used. Casual visits by security firms, which cost less, should be considered. NB: Watchman should not be employed as a matter of course, but only if considered a necessity.

10. Alarms

May be considered necessary, if flimsy temporary building holds very valuable property.

11. Deliveries

(a) Must be physically checked – it is too late to complain afterwards about a mistake sometimes genuine.

(b) Special care must be taken with small items, but bulk supplies such as bricks and ready-mixed concrete still require attention.

(c) Any defects or shortages to be recorded on *both* delivery sheets.

(d) If a reputation is created that everything is checked it has the desired effect.

(e) Delivery notes are *vital*, proof of fraud, payment, etc. Forward to head office *without delay*.

(f) Large sites justify a storeman-cum-checker.

(g) Only named persons of supervisory rank must be allowed to sign for materials.

NB: Sub-contractors must *never* be allowed to sign for receipt of company materials.

12. Plant

Whether owned or hired, all plant and accessories must be used properly and not deliberately or by carelessness or neglect be allowed to become unusable. Costs of repairs, etc. not attributable to fair wear and tear are 'losses'.

NB: *There must be an equal urgency to cancel hire as there is to engage.* Plant not required on site is expensive and is an irrecoverable 'loss'.

Each day before leaving
Check – security of buildings, gates, compound. *All that can be locked up must be.*
Plant – ensure not easily started by intruders.
Even though watchman, etc. is employed, security is still the responsibility of site management.

Index

alarms, types of, 37
arrest, powers of, 53

blocks, storage of, 29
bricks
  common, 29
  facing, 29
  storage of, 29
bungalows, used as stores, 14

car parks, 48
cash on sites,
  site float, 42
  storage, 42
cement, storage of, 30
charges for
  electricity, 53
  gas, 53
checkers
  office location of, 25
  qualifications of, 16
checking
  by cube, 17
  by tare, 17
  of fuel oil, etc., 19, 20
  of quality, 17, 18
clocking-in offences, 51
clothing destroyed by fire, 68
company security procedures
  at pre-contract stage, 8, 13
  at tender stage, 7
  breaches of, 9
  general to avoid losses, 7
  measurements, 6
  ordering, 6
  timekeeping, 6
  unavoidable variations, 8
compensation, 67, 68, 85
compounds
  gates of, 24
  types of, 23, 24

concrete paving slabs, storage of, 30
concrete, ready mixed, checking
  deliveries of, 18
CONSEC, 12, 66, 95
contract period
  hidden costs of extension, 9
  variation orders, 94
courts
  attendance at, 87
  expenses from, 87
  procedure at, 87
Crime Prevention Officers, 13, 14, 81
  85
  see also police
damage
  combatting, 58
  compensation for, 67, 68, 85
  Home Office anti-vandalism report,
    1, 2, 96–8
  pre-start damage, 15
  photographs of, 15
  record of, 15
  types of, 56
dayworks, 71
deliveries
  acceptance of, 16, 18
  checkers, 16
  methods of checking, 16
  notes, 16, 17, 21
  out of hours, 22
  postponing of, 11
  short, 4, 21
  signatures for, 16
  sub-contractors' materials, 22
diary, foreman's use of, 53
discipline procedures
  by warning, etc., 73
  for clocking-in offences, 51
  for theft, 4
dogs, for guarding, 40
doors and door casings

missing, 85, 87
storage of, 31
dumpers, 61
drivers of, 61

electrical goods, storage of, 30
electricity charges, 53
employees
absence of, 76
casual, 3
leaving by, 76
recruitment of, 73
thefts by, 3
empty premises
care of, 84
letter re, 92

fences
entrances, 24
gates in, 24
types of, 23, 24
fire precautions, 49
causes of fires, 49, 50
compensation for clothing destroyed
in site fire, 68
fire service
liaison with, 82
sample letter to, 89
floodlighting, 39
found property on site, 48
fragile materials, storage of, 31
fuel delivery checks, gas, oil, etc., 18,
19, 20

garages, used as stores, 14
gas bottles
security, 19, 20
demurrage charges, 20
gas charges, 53
glass, storage of, 29
guard dogs, 40

holidays, plant security during, 62
huts, on site
bottomless, 25, 27
location of, 25, 26
sectional, 25

keys on site
code system for, 27
holders of, 27
safe custody of, 27

labour relations
discipline, 72

dismissals, 74
employee absent, without excuse, 76
employee leaving, 76
misconduct, 75
periods of notice, 74
reasons for dismissal, 75
recruitment, 72
redundancy, 75
locks
precautions, 26
types of, 25

materials
control of, 10
site issue of, 27
specialised storage, 28–21, 61
subcontractors, 11, 61, 77
mixers, 61

new tenants
liaison with, 84
sample letter to, 91
notices
compound, 24
rewards, 84
site, 25

offcuts, 47
offices on site
location on checkers, 25
pay out, 45
types of, 25
oil storage, 31
order books, temporary, 54
orders, bulk, calling forward, 10
overtime work, 51

packing cases, chargeable, 19
palletts, chargeable, 19
payments
bonus target system, 70
dayworks, 71
form 714 certificate effects on, 45
straw men checks, 71
to subcontractors, 45
wages of absent employee, procedure
for, 46
paving slabs, storage, 30
photographs of pre-start damage, 15
plant
acceptance of, 61
dumpers, 61
holiday periods, 62
identification of, 59
return and collection of, 62

scaffolding, 60
    security of, 59
    subcontractors', 63
plumbing goods, storage, 30
police
    liaison with, 13, 14, 81, 85
    reports to, 85, 87
    sample letter to, 88
    watching site, 37
premises, empty
    sample letter about, 92
    security of, 84
prosecutions, 3, 4, 85
property, found on site, 48
public, general
    liaison with, 83
    sample letter to, 90
purchases, staff, 47

reinforcing wire, storage, 31
restitution, damage and theft, 85
roof trusses, storage, 31
redundancy, 75
rewards, 90
rubbish, trades disposal of, 51

safes, types of, 42
scaffolding, 60
schools
    liaison with, 82
    sample letter to, 89, 90
search, powers of, 53
Security Code of Practice, 100–4
security firms (commercial)
    wages via, 45
    watching by, 35
security
    out of hours, 34
    pre-contract, 8
sheds, security of
    closed site, 27
    open site, 26
shutters, types, 25, 26
staff purchases, 47
staircases, storage and protection of, 31
stock controls, 10
storage, bad, 4
stores, security of
    closed site, 27
    open site, 26

straw men, 77, 78
subbing, 42
sub-contractors
    general control of, 77, 78
    labour only, control of, 79, 80
    materials of, 11, 77
    payments to, 45, 70, 79
    plant, 63, 78
    straw men, 77, 78

telephones
    on site, 52
    payphones, 52
temporary order books, 54
tenants, new
    liaison with, 84
    sample letters to, 91
thefts, procedures, 68, 85, 87
tiles, storage of, 31
timber, storage of, 28
time clocks, 51
trades' rubbish, 51
transfer notes, 22, 52
tools
    marking of, 65
    procedure after theft, 68
    security of, 64
    storage of, 66
    workmen's stolen, compensation, 67

vandalism
    by employees, 5
    by trespassers, 5
    Home Office Committee on, 5
    see also damage
wages
    absent employees, 46
    commercial firms use of, 45
    methods of payment, 44, 45
    pay out office, 45
    subbing, 42
watchmen
    caravans, 35
    commercial security firms, 35
    own employees, 35
    police, 37
window frames, storage of, 29
wire, reinforcing, storage of, 31

youth clubs, liaison with, 83